Springer Tracts in Modern Physics
Volume 198

Managing Editor: G. Höhler, Karlsruhe

Editors: J. Kühn, Karlsruhe
       T. Müller, Karlsruhe
       A. Ruckenstein, New Jersey
       F. Steiner, Ulm
       J. Trümper, Garching
       P. Wölfle, Karlsruhe

Available online
www.springerlink.com/series/stmp/

Now also Available Online

Starting with Volume 165, Springer Tracts in Modern Physics is part of the [SpringerLink] service. For all customers with standing orders for Springer Tracts in Modern Physics, we offer the full text in electronic form via [SpringerLink] free of charge. Please contact your librarian who can receive a password for free access to the full articles by registration at:

www.springerlink.com/orders/index.htm

If you do not have a standing order, you can nevertheless browse through the table of contents of the volumes and the abstracts of each article at:

www.springerlink.com/series/stmp/

There you will also find more information about the series.

Springer
New York
Berlin
Heidelberg
Hong Kong
London
Milan
Paris
Tokyo

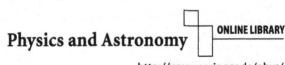

Physics and Astronomy
    ONLINE LIBRARY

http://www.springer.de/phys/

Springer Tracts in Modern Physics

Springer Tracts in Modern Physics provides comprehensive and critical reviews of topics of current interest in physics. The following fields are emphasized: elementary particle physics, solid-state physics, complex systems, and fundamental astrophysics.

Suitable reviews of other fields can also be accepted. The editors encourage prospective authors to correspond with them in advance of submitting an article. For reviews of topics belonging to the above-mentioned fields, they should address the responsible editor, otherwise the managing editor. See also http://www.springer.de/phys/books/stmp.html

Wolfgang Kilian

Electroweak Symmetry Breaking

The Bottom-Up Approach

With 25 Figures

 Springer

Wolfgang Kilian
Theory Group
Deutsches Elektronen-Synchrotron DESY
D-22603 Hamburg
Germany
wolfgang.kilian@desy.de

ISBN 978-1-4419-2310-3   e-ISBN 978-0-387-21635-5

Physics and Astronomy Classification Scheme (PACS): 12.15.Ji, 13.66.Fg, 13.85.Hd, 12.60.-i

Library of Congress Cataloging-in-Publication Data
Kilian, Wolfgang.
   Electroweak symmetry breaking : the bottom-up approach / Wolfgang Kilian.
      p.   cm. — (Springer tracts in modern physics ; 198)
   Includes bibliographical references and index.

   1. Broken symmetry (Physics).   2. Electroweak interactions.   I. Title.   II. Series.
QC793.3.S9K55   2003
539.7'544—dc21   2003050502

Printed in the United States of America.

9 8 7 6 5 4 3 2 1

www.springer-ny.com

Springer-Verlag   New York Berlin Heidelberg
A member of BertelsmannSpringer Science+Business Media GmbH

Preface

Almost 40 years have passed since the theory of Glashow, Salam, and Weinberg explained the weak interactions and unified them with electromagnetism as a quantum field theory. More than 15 years ago the predicted carriers of those interactions, the W and Z vector bosons, were found. Since then, this theory, which has become the Standard Model of particle physics, has thoroughly been tested and verified. The prediction of the top quark mass from precision observables, the consistency of the flavor picture and the prediction of CP violation in the B-meson system are among the important successes. The precision of the data has allowed to check it even at the quantum level. Apart from the recent discovery of neutrino mixing, which can easily be accomodated, no deviation could yet be detected that would make a major revision of the model necessary.

Nevertheless, the central question of the Standard Model has remained open. There is no clear indication of what causes the breaking of the electroweak gauge symmetry, which is the core of the theoretical formalism. While the original idea of an elementary Higgs field with a nontrivial potential is technically a valid explanation of the problem, it is now widely believed that this cannot be the whole answer. The hierarchy problem of the fundamental scales, namely the huge difference between the scales of electroweak symmetry breaking and the scale of gravitation, can apparently be explained only if either the Standard Model in its simplest form is incomplete, or if an alternative mechanism of electroweak symmetry breaking is realized in Nature.

The main obstacle to a solution of this puzzle is the limited energy reach of the existing particle colliders. It has become clear that the natural energy scale of phenomena associated with electroweak symmetry breaking is between 100 GeV and 1 TeV, which has hardly been touched by LEP, SLC, HERA, and Tevatron. At the same time, low-energy experiments have reached an unprecedented precision, and their indirect sensitivity to the structure of electroweak physics complements the direct resolving power of high-energy data, but does not exceed it. Fortunately, a new generation of colliders — the Large Hadron Collider (LHC) and an e^+e^- collider with at least 500 GeV center-of-mass energy — are foreseen to become operational during the next decade. With these machines, it will be possible for the first time to directly probe this energy range, and some of the precision measurements will provide indirect sensitivity up to 10 TeV and beyond.

In view of this situation, the idea of the present book is to summarize the current knowledge about electroweak physics and to implement this in a generic framework which then allows us to identify the observables that are directly associated with the symmetry-breaking sector. For those observables, we discuss the characteristic signatures at the planned colliders, which leads to experimental strategies and finally to estimates for the accuracy in the determination of the relevant parameters, based on realistic Monte Carlo simulations. The effective-theory approach which we will consistently use throughout this work cleanly separates model-independent and model-dependent predictions, where both have to be understood in order to give a meaningful interpretation to the expected experimental data.

The first chapter of this review is devoted to an introduction to the hierarchy problem and to dynamical scale generation as a potential solution, which is behind many directions in model-building. This serves as a background for the later chapters where we discuss the theoretical and experimental ideas of how to address these questions in practice.

In Chapter 2 we will set up the model-independent theoretical framework. Thus, we develop the effective-Lagrangian method for describing electroweak interactions, beginning with the low-energy structure, introducing massive gauge bosons and finally the symmetry-breaking sector. In this *bottom-up* approach, the Standard Model will appear as a special case where the Higgs sector is realized in a specific way, but which otherwise is not singled out as the preferred theory of electroweak symmetry breaking.

To actually probe the symmetry-breaking sector, one needs measurements of Goldstone scattering amplitudes which via the Goldstone-boson equivalence theorem correspond to vector boson scattering amplitudes, the subject of Chapter 3. Throughout the discussion, we consider the various possible realizations of the Higgs sector in parallel. Much of the phenomenology does not depend on whether a physical Higgs particle exists or not. Clearly, the observables which do depend on this alternative are among the ones we are most interested in.

Chapter 4 summarizes the actual prospects and possibilities for measurements at the next generation of colliders. Qualitative conclusions can be drawn by order-of-magnitude estimates using crude approximations, but for a quantitative result one needs precise calculations of the complicated signal and background processes one encounters in Higgs sector physics. Hence, computer tools have been indispensable in arriving at the numerical results that are quoted in this review.

There is an elaborate theory associated with the method of effective Lagrangians. We will not discuss this in detail but refer the reader to the literature [1]. Here, we focus on the particular phenomenology associated with electroweak symmetry breaking, and adopt the language of effective theories as far as it is necessary to precisely define fields, symmetries, and parameters. This suffices to derive scattering amplitudes and observables, and it allows us to restrict the discussion to the physical questions, the structure of the Higgs sector and the mechanism of scale generation. The text is written under the assumption that the reader is familiar with the basics of Lagrangian field theory, Feynman diagrams and collider physics, but the derivations and ideas we discuss do not require a more than superficial knowledge about quantum fields and renormalization theory.

I would like to express my gratitude to H. Anlauf, E. Boos, A. Djouadi, B. Kniehl, M. Krämer, J. Kühn, H.-J. He, S. Heinemeyer, P. Manakos, T. Mannel, G. Moortgat-Pick, M. Mühlleitner, O. Nachtmann, T. Ohl, T. Plehn, A. Pukhov, J. Reuter, K. Riesselmann, M. Spira, C.-P. Yuan, G. Weiglein, and P. Zerwas, as well as to my colleagues in the Physics Department at Karlsruhe for continuing encouragement, valuable discussions and for productive collaborations. Special thanks for generous support go to the Forschergruppe of the Deutsche Forschungsgemeinschaft "Quantenfeldtheorie, Computeralgebra und Monte-Carlo-Simulation", and to the BMBF-Graduiertenkolleg "Elementarteilchenphysik" at the University of Karlsruhe.

Karlsruhe,
January 2003   *Wolfgang Kilian*

Contents

1

Introduction

1.1 The Hierarchy Problem of Particle Physics

As far as we know, the constituents of our world experience four basic types of interaction. These are the electromagnetic, weak, strong, and gravitational forces. Elaborate theories have been developed for all of them in the past century, which are in excellent agreement with all experimental data. The first two are unified in the electroweak theory of Glashow, Salam, and Weinberg [2, 3]. The strong force is described by Quantum Chromodynamics (QCD) [4, 5], and since it shares many properties with the electroweak force, the common description of electroweak and strong interactions is referred to as the Standard Model (SM). Finally, Einstein's theory of General Relativity provides the theory of gravitation, and even though it is not a quantum theory, it is sufficient to describe gravitational interactions as far as they are presently accessible. Nevertheless, there are open questions that cannot be answered by these theories. In this book we will focus on one of them, the *hierarchy problem* of particle-physics scales which is associated with electroweak physics, and discuss how it can be addressed by experiments at present and future colliders.

The strengths of the fundamental forces are characterized by three quantities, namely [6]

$$\text{the proton mass,} \qquad m_p = 0.983\,271\,998(38)\,\text{GeV}, \qquad (1.1)$$

$$\text{the Fermi constant,} \qquad G_F = 1.166\,39(1) \times 10^{-5}\,\text{GeV}^{-2}, \qquad (1.2)$$

$$\text{Newton's constant,} \qquad G_N = 6.707(10) \times 10^{-39}\,\text{GeV}^{-2}, \qquad (1.3)$$

expressed in particle-physics units where $\hbar = c = 1$. Inverting the latter two, we have three different fundamental mass scales,

the proton mass,	m_p	$\approx 1\,\mathrm{GeV}$,	(1.4)
the Fermi scale,	$v = (\sqrt{2}\,G_F)^{-1/2}$	$\approx 246\,\mathrm{GeV}$,	(1.5)
the Planck scale,	$M_{\mathrm{Planck}} = G_N^{-1/2}$	$\approx 1.22 \times 10^{19}\,\mathrm{GeV}$.	(1.6)

All three scales are of immediate relevance for macroscopic physics, even for our everyday life.

First of all, the protons and neutrons which make up atomic nuclei are composite particles, and their masses are determined not by their elementary constituents, but by the inherent scale of the strong interactions which bind them together. Since the dominant contribution to the mass of any atom comes from its nucleus, the strong-interaction scale, represented by the proton mass, sets the *mass scale* of all bodies, from individual atoms up to planets and stars.

According to the SM, the mass scale for elementary particles other than hadrons is set by the Fermi scale. This includes the electron, which happens to be 2000 times lighter than the proton ($m_e = 511\,\mathrm{keV}$) and does not contribute significantly to atomic masses. However, for the very same reason, the *size* of an atom depends mainly on m_e, e.g. the Bohr radius of hydrogen

$$r_\infty \approx \frac{1}{\alpha}\left(\frac{1}{m_e} + \frac{1}{m_p}\right),$$

(1.7)

where α is the electromagnetic fine-structure constant, a dimensionless number. Thus, the electron mass — hence the Fermi scale — sets the *length scale* from atoms up to macroscopic bodies.

The gravitational interaction, summarized in Newton's law

$$F = G_N \frac{m_1 m_2}{r^2},$$

(1.8)

is the only force that cannot be screened. Therefore, it dominates at large distances and it shapes the universe as a whole. While the masses of macroscopic bodies are approximately given by the masses of the atoms of which they consist, hence by the strong-interaction scale, the gravitational force associated to these masses is determined by the Planck scale. The Planck scale sets the overall *weight scale* in macroscopic physics.

Together, the three scales m_p, v, and M_{Planck} provide the dimensions of the world as we know it. (To get a complete picture, some twenty additional *dimensionless* parameters have to be specified.) However, the Planck mass plays a special role. To illustrate this, let us rewrite Newton's law (1.8) in a suggestive form,

$$F = m_1 m_2 \frac{1}{(M_{\mathrm{Planck}} r)^2},$$

(1.9)

which implies that gravitation becomes a strong force for distances of the order of M_{Planck}^{-1}, a factor 10^{19} smaller than the size of an atomic nucleus. Alternatively, we could write

$$F = \left(\frac{m_1}{M_{\text{Planck}}}\right) \left(\frac{m_2}{M_{\text{Planck}}}\right) \frac{1}{r^2}. \tag{1.10}$$

When masses are measured in units of M_{Planck}, gravitation becomes a theory completely free of adjustable parameters.

This simple observation is confirmed by a closer look at the theories involved. The General Theory of Relativity is a theory of space-time geometry and therefore defines the framework for all other interactions. Otherwise, the gravitational interactions it predicts are extremely weak and do not play any role in particle phenomenology. Electroweak and strong interactions are sufficient to explain all observed data. Now, the corresponding renormalizable quantum field theories can in principle be extrapolated down to distances much smaller than their own characteristic scales, $r = m_p^{-1}$ and $r = v^{-1}$. There is no point in considering these scales as fundamental, since even today we can resolve distances smaller than that in collider experiments. However, space resolutions of the order M_{Planck}^{-1} are far beyond experimental reach. A quantum field-theoretical description, as it is realized in the SM, breaks down at this scale. Apparently, the Planck mass sets a fundamental limit, and it may be interpreted as a basic unit of space, time, mass and energy.

Expressed in such "natural" units, the two particle physics scales m_p and v are tiny,

$$m_p/M_{\text{Planck}} \approx 8 \times 10^{-20}, \tag{1.11}$$

$$v/M_{\text{Planck}} \approx 2 \times 10^{-17}. \tag{1.12}$$

Masses are, in some sense, perturbations of the basic equations of motions which are otherwise determined by symmetries. We might simply accept the existence of such small perturbations as we find here. However, if we do not accept it, we have to search for an underlying physical reason which makes the numbers (1.11) and (1.12) as small as they are. Thus, we have identified the *hierarchy problem* [7] of particle physics scales.

1.2 The QCD Solution and the Electroweak Puzzle

The two fundamental theories which make up the SM, the electroweak theory and QCD, are quantum field theories based on the principle of local gauge invariance. A characteristic feature of such theories is the scale dependence of the effective (dimensionless) coupling parameter, which for QCD in leading-logarithmic approximation is given by

$$\alpha_s(\mu) = \frac{4\pi}{7 \ln \frac{\mu^2}{\Lambda^2}}. \tag{1.13}$$

This result takes into account all six quark flavors d, u, s, c, b, t. The constant Λ is a parameter with the dimension of mass whose precise value depends on the renormalization scheme, but which in any scheme is related to the inherent strong-interaction scale which we have identified with the proton mass m_p.

The value $\alpha_s(\mu)$ decreases when the energy scale μ is sent to infinity, a property known as *asymptotic freedom* [5]. Turning this around, if at some high scale $\mu = M_{\text{Planck}}$ an initial value $\alpha_s(M_{\text{Planck}})$ is given, the interaction becomes strong at

$$\Lambda \approx M_{\text{Planck}} \exp\left(-\frac{2\pi/7}{\alpha_s(M_{\text{Planck}})}\right). \tag{1.14}$$

At this low scale, fields condense and bound states like the proton emerge.

Actually, if we knew the theory which unifies the SM with gravitation, we could compute the parameter g_s which appears in the Planck-scale effective Lagrangian of QCD, related to α_s by

$$\alpha_s(\mu) = g_s(\mu)^2/4\pi. \tag{1.15}$$

If this theory gave us a simple order-one number, say $g_s(M_{\text{Planck}}) = 1/2$, we could insert it into (1.14), with a result

$$\Lambda/M_{\text{Planck}} \approx e^{-32\pi^2/7} = 2.5 \times 10^{-20}. \tag{1.16}$$

Incidentally, this is close to the actual value (1.11).

Of course, we have no reason to assume that QCD is valid unmodified up to the Planck scale, and even if it was, we do not know what determines the initial value of g_s. However, this exercise shows that the renormalization group which determines the form of (1.13) typically generates a huge scale ratio if all dimensionless input parameters are of order one. Apparently, in QCD there is no hierarchy problem. Wherever such a mechanism is at work, at the low scale where the coupling becomes strong all effects of a strongly interacting theory should be expected, including field condensation, spontaneous breaking of symmetries, and the emergence of bound states.

While the origin of hadron masses (e.g., proton and neutron) is thus reasonably well understood, the origin of the other elementary particle masses (e.g., leptons and quarks) is unknown. In particular, this applies for the electron mass which determines the size of atoms. In macroscopic terms, we have an idea about the origin of mass, but we do not know the origin of *length*. The electroweak interactions which are felt by the electron do not become strong at the Fermi scale. No sign of electron compositeness has yet been observed.

Instead, a completely different picture emerged when the theory which unifies electromagnetic and weak interactions was found. The relativistic Dirac fields which describe charged fermions can be broken down into two components, left-handed and right-handed chirality states. The two components belong to different representations of the $SU(2)_L \times U(1)_Y$ symmetry group which governs electroweak interactions. For each fermion type, one should therefore list left-handed and right-handed states as independent particle species.

Adopting this point of view, fermion mass terms are caused by bilinear interactions which link different fields. These terms are peculiar since they break the electroweak symmetry. At low energies — below the Fermi scale — the symmetry is hidden. Left- and right-handed states, initially unrelated, mix, and the superpositions can be described in terms of the familiar Dirac fermions. The origin of such bilinear fermion couplings is at the core of the electroweak hierarchy problem.

A similar problem arises for the massive W and Z gauge bosons which mediate weak interactions. The electroweak symmetry requires them to be massless, but they do have masses, of the order of the Fermi scale. This fact can also be attributed to bilinear couplings of two field species in different $SU(2)_L \times U(1)_Y$ representations. In this case, the transversal polarization states of the (massless) gauge bosons are coupled to independent (massless) scalar states. The diagonalization of such a mass matrix results in massive vector bosons as physical eigenstates. Their longitudinal polarization states absorb the scalar degrees of freedom which therefore are unobservable.

Technically, this idea is equivalent to the assumption that the electroweak gauge symmetry is *spontaneously* broken [8] at an energy scale v. With this type of symmetry breaking, the mentioned bilinear couplings can be accomodated. Spontaneous symmetry breaking is always accompanied by massless scalar (Goldstone) bosons which naturally provide the longitudinal components of the massive vector bosons. This effect is known as the *Higgs mechanism*. Goldstone scalars can also be coupled to fermions. The coupling strengths, multiplied by the Fermi scale, are identified with the fermion masses.

Without deeper knowledge about the underlying dynamics the assumption of spontaneous symmetry breaking is a mere mathematical exercise which neither explains fermion and boson masses, nor does it address the electroweak hierarchy problem. The real question is for the physics which is responsible for the Higgs mechanism to occur at such a low scale compared to the fundamental Planck scale. Experiments which uncover the phenomenology associated with the symmetry-breaking sector should give us clues where to look further. The observations may even point to a particular model which solves the problem, as happened with the discovery of QCD jets and scaling, which verified the theory and made the hierarchy problem of hadron masses disappear.

Since we qualitatively understand QCD-like field condensation, many attempts to solve the electroweak hierarchy problem involve a similar mechanism of dynamical symmetry breaking and scale generation. It could either be induced *directly* by some new strong interaction ("technicolor"), leaving hadron-like traces in the high-energy particle spectrum, or *indirectly*. In the latter case, dynamical scale generation would occur in some new sector of particles not directly coupled to the known spectrum. The scale of symmetry breaking would be transferred to the visible world by "messenger" states coupled to both sectors, presumably involving new particles such as a physical Higgs boson. Supersymmetric and "little Higgs" models fall into this category.

Or, the reason for electroweak symmetry breaking could be due to effects beyond the applicability of four-dimensional quantum field theory, e.g., extra dimensions and the appearance of string states.

Without new experimental data available, all such considerations will remain speculative. Detailed explorations of the physics associated with scale generation, the Higgs sector, must be carried out, before anything about the electroweak hierarchy problem can be said with certainty.

2

Effective Theories of Electroweak Interactions

Weak interactions are mediated by massive vector particles, the $W^{\pm}$ and the Z boson. Their discovery was predicted by the model of Glashow, Salam, and Weinberg [2, 3], which provides an accurate description of their properties. It associates electroweak symmetry breaking with a scalar particle, the Higgs boson. It is a renormalizable field theory, therefore in this model (and in many of its extensions), observables can be expressed in terms of a small number of parameters and computed up to high orders in perturbation theory or extrapolated up to energies far beyond the reach of present-day experiments.

While renormalizability is essential for the construction of fundamental quantum field theories, it is not a useful organizing principle for the phenomenological description of physics which has yet to be explored. When adopting a renormalizable field theory, one makes the implicit assumptions that the degrees of freedom included in the Lagrangian (e.g., the Higgs boson whose existence is not even confirmed) exhaust the spectrum, and that these particles are elementary objects. If this is accepted, the number of free parameters of the model is finite. However, the underlying assumptions may be wrong. In fact, asking for the spectrum and for the fundamental degrees of freedom poses the more interesting question, which can only be answered by looking at Nature itself.

Therefore, the framework for understanding physics in an energy range (or at a precision level) that has not been accessed before can only be provided by a *non-renormalizable* field theory, which implements all established facts about low-energy physics, but contains an a priori infinite set of new observable parameters. In some sense, these parameters map the infinite set of possible underlying theories. In the present context, we are interested in the mechanism of electroweak symmetry breaking, so the parameters will be associated with electroweak interactions. Fortunately, it turns out that in the perturbative low-energy and loop expansions the number of parameters which are actually needed remains finite, given the finite accuracy that measurements can deliver. Only when a new particle threshold is reached, does the perturbative expansion break down, but then the new degrees of freedom can directly be observed and subsequently be included in the framework.

This effective-Lagrangian approach has been developed in the context of strong interactions [9] and was adapted later to the low-energy description of electroweak interactions [10]. In the following sections we will review it in some detail, such that the necessary concepts and notations which will be used in the discussion of collider phenomenology in the later chapters of this work can be stated.

2.1 Fermi Theory

Below the threshold where W and Z bosons can be produced directly, light fermions and massless vector bosons (gluons and photons) are the basic degrees of freedom. The effects of massive vector boson exchange are accounted for by non-renormalizable operators in the appropriately constructed effective Lagrangian [11, 12], which generalizes the original Fermi model of weak interactions. This effective theory holds in an energy range between approximately $1\,\mathrm{GeV}$ and $100\,\mathrm{GeV}$. At lower energies QCD confinement makes quarks unobservable. Above this range, the heavy vector bosons become dynamical degrees of freedom which have to be included explicitly in the model.

When constructing the generic effective Lagrangian for the given spectrum, the only restrictions for the form of the operators are set by charge and color conservation. We will ignore gluons for the most part when discussing weak interactions, so the constraint from QCD is that quarks should be coupled as color singlets in the effective Lagrangian. Charge conservation is implemented by the electromagnetic local $U(1)$ invariance of the interactions.

The operators in the effective Lagrangian are naturally ordered by increasing canonical dimension. Hence, we will first discuss the spectrum which is determined by the relevant operators (dimensions two and three), then add the dynamics of electromagnetic interactions which is provided by marginal operators (dimension four), finally the irrelevant operators (dimension five and higher) which include weak interactions [13].

2.1.1 Fermion Masses

Throughout this work, we will use a doublet notation both for left- and right-handed quark and lepton fields,

$$Q_L = \begin{pmatrix} U_L \\ D_L \end{pmatrix}, \qquad\qquad Q_R = \begin{pmatrix} U_R \\ D_R \end{pmatrix}, \qquad (2.1)$$

$$L_L = \begin{pmatrix} N_L \\ E_L \end{pmatrix}, \qquad\qquad L_R = \begin{pmatrix} N_R \\ E_R \end{pmatrix}. \qquad (2.2)$$

All fermion fields are understood as two-component Weyl spinors and carry the additional generation index: $U = u, c, t$ and $D = d, s, b$ for the quarks;

$N = \nu_e, \nu_\mu, \nu_\tau$ and $E = e, \mu, \tau$ for the leptons. The color index of quarks is suppressed.[1]

Since there are pairs of fermion fields of equal charge in the low-energy spectrum, bilinear coupling terms in the Lagrangian are allowed. We define "left-handed" and "right-handed" fields to be in mutually conjugate representations of the Lorentz group. (One can always replace a fermion field by its charge conjugate, such that the chirality of the representation is switched from left-handed to right-handed and vice versa.) Noting that the quarks and the lower leptons are charged and the neutrinos are neutral, charges being equal only for corresponding L and R fermions, the allowed monomials of dimension three read

$$\mathcal{L}_3 = -(\bar{Q}_L M_Q Q_R + \bar{L}_L M_L L_R + \text{h.c.}) - (\bar{N}_L^c M_{N_L} N_L + \bar{N}_R^c M_{N_R} N_R), \quad (2.3)$$

where the masses M_Q, M_L, M_{N_L}, and M_{N_R} are matrices in generation space. The second bracket can also be written using a projection operator,

$$\mathcal{L}_{3(N)} = -\left(\bar{L}_L^c M_{N_L} \frac{1+\tau^3}{2} L_L + \bar{L}_R^c M_{N_R} \frac{1+\tau^3}{2} L_R \right). \quad (2.4)$$

where $\tau^{1,2,3}$ are the Pauli matrices. Charge conservation forbids self-coupling terms such as $\bar{Q}_L^c Q_L$. Therefore, the mass matrix for any charged fermion is degenerate, e.g. for the electron

$$\mathcal{L}_{3(e)} = -(\bar{e}_L \ \bar{e}_R^c) \begin{pmatrix} 0 & m_e \\ m_e & 0 \end{pmatrix} \begin{pmatrix} e_L^c \\ e_R \end{pmatrix} \quad (2.5)$$

it has two eigenvalues $\pm m_e$, with equal mixtures of the spin-$\frac{1}{2}$ fields e_L and e_R states being the eigenvectors. This makes up four states for a physical electron, a Dirac particle.

Since the fermion mass terms are the operators with lowest dimension in the effective Lagrangian, they are the most important ones for energies going to zero, distances going to infinity. Physical particle states coincide with *eigenstates* of the mass matrices. Thus, if we adopt the basis of asymptotic (classical) states, the mass matrices M_Q and M_L are diagonal. The eigenvalues are the physical masses of the particles as indicated in the following graph:

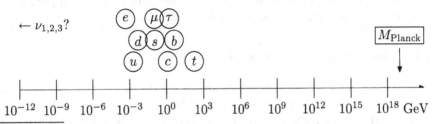

---

[1] For simplicity, we include the top quark in the low-energy effective theory, even though it is heavier than the W and Z bosons. The construction of an effective theory where the top quark is properly integrated out is possible, but introduces unnecessary complications.

On a logarithmic scale the charged fermion masses are roughly uniformly distributed between 10^{-3} and 10^3 GeV.

With the charge conservation constraint being absent, for the neutrinos there is no reason to expect a degeneracy in their mass matrices which would make them Dirac fermions.[2] Assuming that there are three right-handed neutrinos, the neutrino mass term reads

$$\mathcal{L}_{3(N)} = -(\bar{N}_L \ \bar{N}_R^c) \begin{pmatrix} M_{N_L} & M_L \\ M_L^T & M_{N_R} \end{pmatrix} \begin{pmatrix} N_L^c \\ N_R \end{pmatrix}, \qquad (2.6)$$

so the neutrino mass matrix has six different eigenvalues, corresponding to six Majorana-type eigenstates with two degrees of freedom each. Again, the matrix is diagonal in the basis of asymptotic states, but this time the eigenvectors are in general not equal mixtures of N_L and N_R.

From the LEP 1 data we know that three eigenvalues of the neutrino mass matrix are smaller than 45 GeV. In fact, these are much smaller than the mass of any other fermion. Only upper bounds for the individual (left-handed) neutrino masses have been established ($m_{\nu_e} < 3$ eV, $m_{\nu_\mu} < 190$ keV, $m_{\nu_\tau} < 18$ MeV), and only differences of squared masses could actually be measured (considerably less than 1 eV^2) [6]. Concerning the right-handed states, not much is known with certainty. Masses of right-handed neutrinos are essentially unconstrained from experimental data since they do not feel any known interaction except for gravitation. While the numbers of quark, charged lepton, and left-handed neutrino species are related by the requirement of anomaly cancellation in the electroweak theory, no such constraint is known for right-handed neutrinos.

Parity is identified as an accidental low-energy symmetry. Because charge and color conservation forbid diagonal terms in the mass matrices for charged leptons and quarks, a discrete symmetry emerges which exchanges left- and right-handed fermions. The non-relativistic states are equal superpositions of both. This is true for any charged fermion. In conjunction with the classical limit of the electromagnetic field and a convention for assigning charges and magnetic moment directions, one can assign helicity quantum numbers to the individual components. However, at higher energies where it becomes apparent that "left"- and "right"-handed fields are in fact unrelated (having different weak interactions), parity is no longer a symmetry.

2.1.2 Electromagnetic Interactions

Marginal (dimension four) operators are necessary to describe propagating fields and their long-range interactions. For the fermions, these are the kinetic energy and the interaction with the electromagnetic field A^μ, expressed in terms of the covariant derivative

---

[2] If neutrinos are Dirac particles, which means $M_{N_L} = M_{N_R} = 0$, there is an unbroken global $U(1)$ symmetry in the interactions.

$$D_\mu = \partial_\mu + ieqA_\mu, \qquad (2.7)$$

where e is the positron charge and q the multiplying factor for the charge of a given fermion species.

Together with the kinetic energy of the electromagnetic field, expressed in terms of the electromagnetic field strength tensor

$$A_{\mu\nu} = \partial_\mu A_\nu - \partial_\nu A_\mu, \qquad (2.8)$$

the possible gauge-invariant dimension-four operators in the effective Lagrangian are

$$\mathcal{L}_4 = \bar{Q}_L i \not{D} Q_L + \bar{Q}_R i \not{D} Q_R + \bar{L}_L i \not{D} L_L + \bar{L}_R i \not{D} L_R - \tfrac{1}{4} A_{\mu\nu} A^{\mu\nu}. \qquad (2.9)$$

In the doublet notation used here, the normalized charge q is a diagonal 2×2 matrix which reads

$$q_Q = \frac{y_Q + \tau^3}{2}, \qquad q_L = \frac{y_L + \tau^3}{2}, \qquad (2.10)$$

for the quarks and leptons, respectively, with

$$y_Q = \tfrac{1}{3} \quad \text{and} \quad y_L = -1. \qquad (2.11)$$

In addition to the electromagnetic local $U(1)$ invariance, the Lagrangian consisting of $\mathcal{L}_3$ and $\mathcal{L}_4$ only has global symmetries. These correspond to the exact conservation of fermion number (i.e., a global $U(1)$ symmetry) for each quark and charged lepton species, and the conservation of fermion number modulo 2 for each neutrino species. All terms allowed by these low-energy symmetries are present in the effective Lagrangian.

2.1.3 Dimension-Five Operators

At the next order in the low-energy expansion, magnetic-moment type operators (dimension five) appear:

$$\mathcal{L}_5 = \bar{Q}_R \mu_Q \sigma_{\mu\nu} A^{\mu\nu} Q_L + \bar{L}_R \mu_L \sigma_{\mu\nu} A^{\mu\nu} L_L + \text{h.c.} \qquad (2.12)$$

In general, the couplings μ_Q and μ_L are 6×6 matrices in flavor space. They have mass dimension -1, hence their effects are suppressed by a factor E/Λ, where E is the highest energy, momentum, or mass scale in a given experiment, and $\Lambda \sim \mu^{-1}$ is the mass scale set by the operator coefficients. In the effective action, such operators are induced by radiative corrections from the electromagnetic interaction Lagrangian $\mathcal{L}_4$. An example is the one-loop correction to the electron g factor which induces an anomalous magnetic moment in the non-relativistic limit.

If radiative corrections involving the electromagnetic interactions were the only source of $\mathcal{L}_5$ terms in the effective action, they would exhibit the same symmetries as $\mathcal{L}_3 + \mathcal{L}_4$, so the coefficient matrices would be diagonal. This is not the case: A notable example is the operator $\bar{s}\sigma_{\mu\nu}A^{\mu\nu}b$ which leads to the quark decay $b \rightarrow s\gamma$. These *flavor-changing neutral currents* can be explained as loop corrections involving weak interactions, which are discussed in the following section. It might be that there are additional effects at this order that cannot be accounted for by electromagnetic and weak interactions, but the current precision of flavor-physics experiments and calculations is not yet sufficient to answer this question.

2.1.4 Current-Current Interactions

Weak interactions are mediated by dimension-six operators. They break the flavor symmetries of electromagnetic interactions. This is the reason why they are important — the source of nuclear fusion and radioactive decay — despite the fact that they are formally irrelevant at low energies.

A completely generic four-fermion operator (dimension-six) corresponds to an expression like

$$\mathcal{L}_6 = \sum_{f=Q_L,Q_R,L_L,L_R} s_{ijkl}(\bar{f}_i f_j)(\bar{f}_k f_l) + v_{ijkl}(\bar{f}_i \gamma^\mu f_j)(f_k \gamma_\mu f_l) \qquad (2.13)$$

with arbitrary coupling matrices s_{ijkl} and v_{ijkl}, where the brackets enclose color-singlets in the case of quarks. However, from flavor physics data we know that the structure of the weak-interaction Lagrangian is much more restricted. To lowest order, the four-fermion operators that mediate low-energy lepton and quark interactions can be written as a sum of products of charged and neutral currents

$$\mathcal{L}_6 = -4\sqrt{2}\, G_F \left(2J_\mu^+ J^{\mu-} + \rho_* J_\mu^0 J^{\mu 0} \right), \qquad (2.14)$$

which are given by

$$J_\mu^\pm = \frac{1}{\sqrt{2}}\left[\bar{Q}_L \tau^\pm \gamma_\mu V_{\mathrm{CKM}}^{(\dagger)} Q_L + \bar{L}_L \tau^\pm \gamma_\mu L_L \right], \qquad (2.15)$$

$$J_\mu^0 = \left[\bar{Q}_L \left(-q_Q s_w^2 + \frac{\tau^3}{2} \right) \gamma_\mu Q_L + \bar{Q}_R(-q_Q s_w^2)\gamma_\mu Q_R \right.$$
$$\left. + \bar{L}_L \left(-q_L s_w^2 + \frac{\tau^3}{2} \right) \gamma_\mu L_L + \bar{L}_R(-q_L s_w^2)\gamma_\mu L_R \right]. \qquad (2.16)$$

In the leptonic part there are only three parameters. The Fermi constant G_F determines the strength of the charged-current interactions. The neutral-current interactions have been measured to be of the same strength, i.e., $\rho_* \approx 1$. The ratio of the isospin and electromagnetic components of the neutral current is given by s_w, the sine of the weak mixing angle. In the hadronic part, V_{CKM} is the 3×3 quark mixing matrix introduced by Cabbibo, Kobayashi, and Maskawa [14]. We will assume that V_{CKM} is a unitary matrix, as suggested by direct measurements, by the smallness of (loop-induced) flavor-changing neutral currents, and by the observed pattern of CP violation in the flavor sector [6].

In the bottom-up approach pursued here, we may consider the parameters e, s_w, ρ_*, and G_F as input parameters which are measured in specific (low-energy) experiments. When using their values for computing different (in particular, high-energy) processes, it is mandatory to include quantum corrections, the dominant part of which can be absorbed in scale-dependent running couplings. Since some of these corrections (in particular, the scale dependence of the effective electromagnetic coupling) depend on nonperturbative physics, it is often more convenient to use a set of observables defined in high-energy experiments, e.g., on the Z pole. However, a discussion of these effects is beyond the scope of this review (see, e.g., [15, 16]), and we will implicitly assume that an appropriate scheme has been adopted and the computable radiative corrections are included when necessary in practical applications.

Comparing (2.14) with the generic form (2.13), the number of free parameters in the Fermi Lagrangian is much less than allowed by charge and lepton family number conservation. The particular current-current structure of this interaction indicates the presence of an underlying symmetry. This idea led to the Glashow-Salam-Weinberg theory [2], now known as the Standard Model of electroweak interactions.

2.1.5 Symmetries of the Weak Interactions

The local symmetry of the weak interaction Lagrangian $\mathcal{L}_6$ (2.14) is made manifest by introducing vector fields $W^\pm$ and Z,

$$\mathcal{L}_6 = -g_W(W^{+\mu}J_\mu^+ + W^{-\mu}J_\mu^-) - g_Z(Z^\mu J_\mu^0) + \mathcal{L}_{2(W)}, \tag{2.17}$$

with

$$\mathcal{L}_{2(W)} = M_W^2 W^{+\mu} W_\mu^- + \tfrac{1}{2} M_Z^2 Z^\mu Z_\mu. \tag{2.18}$$

Since at low energies the weak-interaction gauge bosons are unobservable, in the Fermi theory they are auxiliary fields (i.e., without kinetic terms), and the field equations which follow from (2.17) do not contain derivatives. They can be solved algebraically,

$$W_\mu^\pm = \frac{g_W}{M_W^2} J_\mu^\mp, \qquad Z_\mu = \frac{g_Z}{M_Z^2} J_\mu^0, \tag{2.19}$$

and inserted back in (2.17), to recover the original form (2.14). This works if the coupling parameters g_W, g_Z and the mass parameters M_W, M_Z are related to each other

$$g_W^2 = \frac{4M_W^2}{v^2} \quad \text{and} \quad g_Z^2 = \rho_* \frac{4M_Z^2}{v^2}. \tag{2.20}$$

We have introduced the electroweak scale v which is related to the Fermi constant by

$$v \equiv (\sqrt{2} G_F)^{-1/2}. \tag{2.21}$$

As long as the W and Z fields do not correspond to physical degrees of freedom, the values of g_W and g_Z (or M_W and M_Z) are arbitrary.

If there is a local gauge symmetry, the vector fields are the affine connections that couple to the fermion fields via covariant derivatives. For left-handed and right-handed fermions, these are

$$D_{L\mu} = \partial_\mu + \mathrm{i}eqA_\mu + \mathrm{i}g_Z \left(-qs_w^2 + \frac{\tau^3}{2} \right) Z_\mu$$
$$+ \mathrm{i}\frac{g_W}{\sqrt{2}} \left(\tau^+ W_\mu^+ V_{\mathrm{CKM}}^\dagger + \tau^- W_\mu^- V_{\mathrm{CKM}} \right), \quad \text{and} \tag{2.22}$$

$$D_{R\mu} = \partial_\mu + \mathrm{i}eqA_\mu + \mathrm{i}g_Z(-qs_w^2)Z_\mu, \tag{2.23}$$

where the CKM matrix is present for quarks only. The Lagrangian of electromagnetic and weak interactions (2.9, 2.17) then reads

$$\mathcal{L} = \bar{Q}_L\mathrm{i}\slashed{D}_L Q_L + \bar{Q}_R\mathrm{i}\slashed{D}_R Q_R + \bar{L}_L\mathrm{i}\slashed{D}_L L_L + \bar{L}_R\mathrm{i}\slashed{D}_R L_R - \tfrac{1}{4}A_{\mu\nu}A^{\mu\nu} + \mathcal{L}_{2(W)} + \mathcal{L}_3 \tag{2.24}$$

where the mass terms $\mathcal{L}_{2(W)}$ and $\mathcal{L}_3$ are given by (2.18) and (2.3), respectively.

If the expressions (2.22, 2.23) are indeed covariant derivatives, one has to identify the corresponding gauge symmetry. The algebra of the charged currents closes with a neutral current proportional to $[\tau^+, \tau^-] = \tau^3$. Correspondingly, the vector fields W^+ and W^- or, equivalently, W^1 and W^2 with

$$W^+ = \frac{1}{\sqrt{2}}(W^1 - \mathrm{i}W^2), \qquad W^1 = \frac{1}{\sqrt{2}}(W^+ + W^-), \tag{2.25}$$

$$W^- = \frac{1}{\sqrt{2}}(W^1 + \mathrm{i}W^2), \qquad W^2 = \frac{\mathrm{i}}{\sqrt{2}}(W^+ - W^-), \tag{2.26}$$

cannot be a doublet but there must be a neutral third component W^3 with a coupling equal to $g_W\tau^3/2$. Thus, the local gauge symmetry contains a $SU(2)_L$ group, where the left-handed fields transform as doublets and the right-handed fields as singlets.

The neutral-current part of the Fermi interaction is not just proportional to τ^3, therefore the Z field must be a linear combination of W^3 and another neutral field B. Assuming that the charged currents are purely left-handed, a single pair of charged $W^\pm$ fields accounts for all all flavor-changing effects. Then, there is no room for additional non-abelian symmetries in the low-energy effective theory, and B should be associated to the generator of a $U(1)$ symmetry. Furthermore, the $SU(2)_L$ algebra forbids any couplings of the W^3 field not proportional to τ^3. If no extra $U(1)$ factors are to be introduced, this is possible only if the photon field A is another linear combination of

W^3 and B such that the W^3 couplings proportional to q cancel. Combining thus all facts known about charged- and neutral-current interactions, even without introducing propagating vector bosons one can identify the familiar $SU(2)_L \times U(1)_Y$ group as an underlying local symmetry. The actual symmetry group of the weak interactions may be larger, but this is not required by the structure of the Fermi Lagrangian.

In the standard derivation of the electroweak theory, the physical Z and A fields are rotated into the gauge fields W^3 and B according to

$$Z = c_w W^3 - s_w B, \qquad\qquad W^3 = c_w Z + s_w A, \qquad (2.27)$$
$$A = s_w W^3 + c_w B, \qquad\qquad B = -s_w Z + c_w A, \qquad (2.28)$$

where $c_w = \sqrt{1 - s_w^2}$, and s_w is the weak mixing angle as measured in neutral-current interactions. However, this particular field redefinition is not enforced by the underlying symmetry [16]. Instead, we should make the general ansatz

$$Z = c_w(1 + x_Z)W^3 - s_w(1 + y_Z)B, \qquad (2.29)$$
$$A = s_w(1 + x_A)W^3 + c_w(1 + y_A)B, \qquad (2.30)$$

with four real parameters x_Z, y_Z, x_A, y_A. Anticipating the fact that they are small, we will work only to first order in these parameters.

The mixing parameters are not completely arbitrary. For a $SU(2)_L \times U(1)_Y$ gauge symmetry, we must be able to rewrite the covariant derivatives (2.22, 2.23) as

$$D_{L\mu} = \partial_\mu + i\left(g_1' q - g_2' \frac{\tau^3}{2}\right)B_\mu + ig W_\mu^a \frac{\tau^a}{2}, \qquad (2.31)$$

$$D_{R\mu} = \partial_\mu + i\left(g_3' q - g_4' \frac{\tau^3}{2}\right)B_\mu, \qquad (2.32)$$

with some coupling constants g, g_i' $(i = 1, 2, 3, 4)$.

The (leptonic) charged current is not modified in (2.31) compared to (2.22), therefore

$$g_W = g. \qquad (2.33)$$

The hadronic charged current is diagonalized by independent left-handed and right-handed rotations of the quarks.

The $SU(2)_L$ symmetry then determines the neutral-current coupling proportional to τ^3 and forbids right-handed W^3 couplings, hence

$$g_Z = \frac{1}{c_w}(1 - x_Z)g, \qquad (2.34)$$

$$g = \frac{e}{s_w}(1 + x_A). \qquad (2.35)$$

Finally, the symmetry should be consistent at the quantum level. Summing over complete generations of fermions, the gauge anomalies cancel if there is a single hypercharge coupling g',

$$g_i' = g', \quad i = 1, 2, 3, 4, \tag{2.36}$$

which implies

$$y_A + x_Z = x_A + y_Z \tag{2.37}$$

and

$$g' = \frac{e}{c_w}(1 + y_A). \tag{2.38}$$

With these couplings, the covariant derivatives are

$$D_{L\mu} = \partial_\mu + ig'\left(q - \frac{\tau^3}{2}\right) B_\mu + igW_\mu^a \frac{\tau^a}{2}, \tag{2.39}$$

$$D_{R\mu} = \partial_\mu + ig'qB_\mu, \tag{2.40}$$

and the fermion fields transform under local $SU(2)_L$ transformations,

$$U(x) = \exp(-i\alpha^a(x)\,\tau^a/2), \quad a = 1, 2, 3, \tag{2.41}$$

as

$$L_L \to UL_L, \qquad\qquad Q_L \to UQ_L, \tag{2.42}$$

$$L_R \to L_R, \qquad\qquad Q_R \to Q_R, \tag{2.43}$$

and under $U(1)_Y$ transformations,

$$V(x) = \exp(i\beta(x)\,\tau^3/2), \tag{2.44}$$

as

$$L_L \to \exp(-i\beta q_L)\,VL_L, \qquad Q_L \to \exp(-i\beta q_Q)\,VQ_L, \tag{2.45}$$

$$L_R \to \exp(-i\beta q_L)\,L_R, \qquad Q_R \to \exp(-i\beta q_Q)\,Q_R, \tag{2.46}$$

where $q_{L,R}$ are the diagonal charge matrices (2.10).

To summarize, the mixing parameters x_Z, y_Z, x_A, y_A determine the gauge couplings g, g' and g_W, g_Z (or the mass parameters M_W and M_Z, cf. (2.20)), once the observables e, s_w, ρ_*, and G_F are fixed. Turning this argument around, it is not possible to predict the gauge boson masses and couplings from low-energy data, since there are free parameters in the low-energy realization of the symmetry which do not correspond to observables. In view of this, the discovery of the W and Z bosons was a nontrivial success of the model of Glashow, Salam, and Weinberg with a minimal Higgs sector where, to leading order, x_Z, y_Z, x_A, y_A all vanish.

2.2 Massive Gauge Bosons

Since the W and Z bosons have been discovered and their properties have been thoroughly explored in the LEP, SLC, HERA, and Tevatron experiments, they are naturally identified with the gauge fields in the auxiliary-field formalism.[3] The mass parameters in the auxiliary-field formulation are then given by the physical W and Z masses,

$$M_W = 80.4\,\text{GeV}, \quad M_Z = 91.2\,\text{GeV}, \tag{2.47}$$

and the gauge couplings g and g' can consistently be extracted from data in many production and decay channels of the W and Z bosons [17, 6]. To make vector bosons physical degrees of freedom, we have to add kinetic terms for the vector fields in the effective electroweak Lagrangian which is valid at and above the energy scale $E \sim M_W \sim M_Z$.

Ignoring anomalous couplings for the moment, these are

$$\mathcal{L}_{4(W)} = -\tfrac{1}{2}\,\text{tr}\,[\mathbf{W}_{\mu\nu}\mathbf{W}^{\mu\nu}] - \tfrac{1}{2}\,\text{tr}\,[\mathbf{B}_{\mu\nu}\mathbf{B}^{\mu\nu}], \tag{2.48}$$

where the field strength tensors are defined in terms of W_μ^a $(a = 1, 2, 3)$ and B_μ as

$$\mathbf{W}_{\mu\nu} = \partial_\mu \mathbf{W}_\nu - \partial_\nu \mathbf{W}_\mu + ig[\mathbf{W}_\mu, \mathbf{W}_\nu], \tag{2.49}$$

$$\mathbf{B}_{\mu\nu} = \partial_\mu \mathbf{B}_\nu - \partial_\nu \mathbf{B}_\mu, \tag{2.50}$$

with

$$\mathbf{W}_\mu = W_\mu^a \frac{\tau^a}{2} \quad \text{and} \quad \mathbf{B}_\mu = B_\mu \frac{\tau^3}{2}. \tag{2.51}$$

The rotation matrices of the physical fields have been introduced above (2.25–2.28).

The transition to the low-energy effective theory is a smooth one. At energies below the W mass the derivatives in (2.49, 2.50) which become energy/momentum factors in the matrix elements are small, allowing for an expansion in powers of E/M_W. To leading order where the momentum factors are neglected altogether, Fermi theory (in the auxiliary-field formulation) is recovered. In loop diagrams, the truncation of the expansion leads to additional divergences in the low-energy effective theory which can be removed by adding matching corrections. In principle, the non-abelian structure in (2.49) gets translated into six- and eight-fermion operators in the low-energy effective Lagrangian which extend the current-current interaction pattern, but they are difficult to access.

2.2.1 Symmetrizing the Lagrangian

While the dimension-four part of the effective Lagrangian (2.24, 2.48) exhibits a manifest gauge symmetry, this symmetry is apparently not present in

---

[3] In the presence of an enlarged gauge group, this may not be entirely correct. However, if any extra gauge bosons are considerably heavier than W and Z, there is an intermediate effective theory where only the latter are present, and the effects of extra massive degrees of freedom can completely be absorbed in $SU(2)_L \times U(1)_Y$-invariant higher-dimensional operators as discussed below.

the dimension-two and dimension-three operators (2.18, 2.3), the fermion and gauge boson mass terms. However, this problem can formally be solved without losing the universality of the effective-theory formalism by introducing an extra field Σ with a suitable transformation law [9, 10, 18].[4] The low-energy spectrum and interactions remain identical to the original model if one assumes that the symmetry is spontaneously broken, as will be discussed in the following section. Thus, the *presence* of the field Σ in the Lagrangian does not affect phenomenology in any way, and there is no additional information contained in the modified Lagrangian. On the other hand, the number of degrees of freedom used for a *parameterization* of Σ has distinct observable consequences which we will investigate in later sections.

The field $\Sigma(x)$ is a 2×2 matrix which transforms under local $SU(2)_L$ transformations $U(x)$ (2.41) and $U(1)_Y$ transformations $V(x)$ (2.44) as

$$\Sigma(x) \rightarrow U(x)\,\Sigma(x)\,V^\dagger(x). \tag{2.52}$$

This transformation property allows us to cancel the transformation matrices of the fermions and vector bosons by inserting extra Σ factors in the non-invariant operators (2.3) and (2.18).

The fermion mass term (2.3) in the effective Lagrangian is replaced by

$$\begin{aligned}
\mathcal{L}_3 = &-(\bar{Q}_L \Sigma M_Q Q_R + \bar{L}_L \Sigma M_L L_R + \text{h.c.}) \\
&- \bar{L}_L^c \Sigma^* M_{N_L} \frac{1+\tau^3}{2} \Sigma L_L - \bar{L}_R^c M_{N_R} \frac{1+\tau^3}{2} L_R,
\end{aligned} \tag{2.53}$$

which has the required $SU(2)_L \times U(1)_Y$ symmetry.

The boson mass term (2.18) is replaced by a kinetic-energy term for the Σ field. We introduce two further abbreviations

$$V_\mu = \Sigma (D_\mu \Sigma)^\dagger \quad \text{and} \quad T = \Sigma \tau^3 \Sigma^\dagger, \tag{2.54}$$

where the covariant derivative is determined by the transformation law (2.52),

$$D_\mu \Sigma = \partial_\mu \Sigma + ig\mathbf{W}_\mu \Sigma - ig' \Sigma \mathbf{B}_\mu, \tag{2.55}$$

to write this as

$$\mathcal{L}_{2(W)} = -\frac{v^2}{4} \operatorname{tr}[V_\mu V^\mu] - \beta' \frac{v^2}{8} \operatorname{tr}[T V_\mu] \operatorname{tr}[T V^\mu] \tag{2.56}$$

with a free parameter β'.

The dimension-four part of the Lagrangian does not necessarily involve the Σ field,

---

[4] Strictly speaking, for a model-independent analysis, one can work with the non-symmetric Lagrangian without loss of generality [19], but then the freedom of parameterizing the Higgs sector and the relations of vector bosons and Goldstone interactions are lost.

$$\mathcal{L}_4 = \bar{Q}_L i\slashed{D} Q_L + \bar{Q}_R i\slashed{D} Q_R + \bar{L}_L i\slashed{D} L_L + \bar{L}_R i\slashed{D} L_R$$
$$- \tfrac{1}{2} \operatorname{tr}\left[\mathbf{W}_{\mu\nu}\mathbf{W}^{\mu\nu}\right] - \tfrac{1}{2} \operatorname{tr}\left[\mathbf{B}_{\mu\nu}\mathbf{B}^{\mu\nu}\right], \tag{2.57}$$

but there is now the possibility to add an invariant potential term

$$\mathcal{L}_\Sigma = -\frac{\mu^2 v^2}{4} \operatorname{tr}\left[\Sigma^\dagger \Sigma\right] + \frac{\lambda v^4}{16} \operatorname{tr}\left[\Sigma^\dagger \Sigma\right]^2 + \cdots. \tag{2.58}$$

The effective Lagrangian constructed this way,

$$\mathcal{L} = \mathcal{L}_{2(W)} + \mathcal{L}_3 + \mathcal{L}_4 + \mathcal{L}_\Sigma, \tag{2.59}$$

is manifestly invariant under the full electroweak symmetry group.

2.2.2 Spontaneous Symmetry Breaking

When constructing a quantum field theory in the Lagrangian formalism, one has to specify the (perturbative) ground state of the system. This ground state determines the physical mass spectrum. If the interactions (2.53, 2.56) are to provide masses for the fermions and vector bosons, we have to assume that the (invariant) operator $\operatorname{tr}\left[\Sigma^\dagger \Sigma\right]$ has a nonzero vacuum expectation value. This may be the result of strong interactions which are beyond the applicability of perturbation theory (cf. the gluon condensate in QCD), but phenomenologically it can be realized as a nontrivial classical minimum of the potential (2.58). We normalize our units such that this expectation value is equal to unity:

$$\langle \tfrac{1}{2} \operatorname{tr}\left[\Sigma^\dagger(x)\Sigma(x)\right]\rangle = 1. \tag{2.60}$$

The Lagrangian is invariant under a group of transformations that change the Σ field but not $\operatorname{tr}\left[\Sigma^\dagger \Sigma\right]$, so in the classical theory there is an infinite degeneracy of equivalent ground states. In quantum field theory, the Hilbert space in this case breaks down into unitarily inequivalent superselection sectors which can be characterized by the average value of the field at infinity [20, 1]. To make up the physical spectrum, one can choose any one of those. It is clearly convenient to take the one where

$$\langle \Sigma(x)\rangle \to 1 \quad \text{for } x \to \infty. \tag{2.61}$$

Thus, the global part of the electroweak symmetry group (the group of transformations at infinity) is spontaneously broken down to the subgroup which leaves $\langle \Sigma\rangle$ invariant, the electromagnetic $U(1)_\mathrm{em}$ symmetry. The local symmetry is manifest, but as far as local transformations beyond $U(1)_\mathrm{em}$ are concerned, it is restricted to transformations that approach the identity for $x \to \infty$.

For perturbative calculations in quantum field theory, this symmetry must be broken by hand, selecting one particular field configuration as the starting point. This requires the introduction of gauge-fixing and ghost terms (cf.

Section 2.3.1). Here we just note that, at least for lowest-order computations, one can choose the *unitary gauge* where

$$\Sigma(x) \equiv 1. \tag{2.62}$$

In this gauge, the effective Lagrangian (2.59) coincides with the non-invariant effective Lagrangian described before where the Σ field was absent. For instance, the composite vector field V_μ (2.54) in unitary gauge is a linear combination of W and Z fields (but no contribution of the photon),

$$
\begin{aligned}
V_\mu &= -ig\mathbf{W}_\mu + ig'B_\mu \\
&= -i\frac{g_W}{\sqrt{2}}(W_\mu^+\tau^+ + W_\mu^-\tau^-) - ig_Z Z_\mu\frac{\tau^3}{2},
\end{aligned}
\tag{2.63}
$$

where the vector boson couplings g_W and g_Z have been defined in Section 2.1.5.

Expanding (2.56) in unitary gauge, we obtain

$$\mathcal{L}_{2(W)} = \frac{g_W^2 v^2}{2} W_\mu^+ W^{-\mu} + \frac{g_Z^2 v^2}{4}(1 + \beta')Z_\mu Z^\mu, \tag{2.64}$$

and can read off the vector boson masses,

$$M_W = \tfrac{1}{2}g_W v \quad \text{and} \quad M_Z = \tfrac{1}{2}g_Z(1 + \beta'/2)v. \tag{2.65}$$

The ρ_* parameter of the neutral-current interactions (2.14, 2.20) is thus given by

$$\rho_* = 1 - \beta', \tag{2.66}$$

and $\rho_* = 1$ implies $\beta' = 0$.

The left-handed (Majorana) neutrino mass term in (2.53) contains two factors of the Σ field while the right-handed neutrino mass term is consistent with electroweak symmetry without extra Σ factors. The couplings of operators with Σ factors which have the dimension of mass are constrained by the cutoff in the low-energy effective theory (cf. Section 3.1.5) and cannot arbitrarily exceed the electroweak scale v. However, there is nothing that prevents M_{N_R} from acquiring a large value, such that the right-handed neutrinos disappear from the low-energy spectrum. At energies below that scale, one can impose lepton number conservation on the effective Lagrangian, which implies $M_{N_L} = 0$. This symmetry will be broken explicitly by the virtual effects of the right-handed neutrinos which are integrated out, but the resulting coefficient matrix M_{N_L} is naturally small, given by the *see-saw* formula [21],

$$M_{N_L} = M_L^T(M_R)^{-1}M_L. \tag{2.67}$$

This scenario is supported by the experimental fact that the (left-handed) neutrino masses and mixing parameters are nonvanishing, but orders of magnitude smaller than the charged lepton masses.

2.2.3 The Higgs Boson

There is one new feature in the spontaneously broken model. Once we have introduced the field Σ, we have to take care of fluctuations of the symmetric composite field $\mathrm{tr}\left[\Sigma^\dagger \Sigma\right]$. This is an additional degree of freedom which can be associated to a particle, the Higgs boson [8]. There are two extreme possibilities:

1. This field is a mathematical entity which does not provide an observable asymptotic state. This happens if the effective mass of such fluctuations is above the energy threshold where the effective theory ceases to be valid. One would expect a nontrivial spectrum of new states in this energy range, and the Higgs state need not play a prominent role [22].
2. The effective mass of fluctuations is of the same order (or smaller) as the W and Z masses. In this case, a Higgs boson should be explicitly included in the spectrum [2].

Clearly, intermediate cases are possible as well.

In addition, there may be other new states in the physical spectrum, not necessarily related to the Higgs sector, which appear at energies around or above the electroweak scale v. If there is a finite number of them, they can be assigned field operators and be coupled to the fields present in the effective Lagrangian. On the other hand, if the number of degrees of freedom is infinite or if no sensible single-particle states can be constructed, the effective-theory description breaks down and should give way to a more fundamental theory valid at these scales.

2.2.4 Nonlinear Symmetry Representation

To realize the Higgs-less case in perturbation theory, one can impose the relation

$$\Sigma^\dagger \Sigma \equiv 1, \tag{2.68}$$

which is consistent with $SU(2)_L \times U(1)_Y$ symmetry, cf. (2.52). This is a stronger statement than (2.60), which explicitly forbids fluctuations. It makes Σ a unitary matrix. Since the gauge transformations of Σ (2.52) span the whole special unitary group, no further constraints on Σ are possible. A unitary 2×2 matrix has three degrees of freedom, therefore the minimal number of fields parameterizing Σ is three. A possible parameterization is given by [9, 10, 18]

$$\Sigma(x) = \exp\left(-\frac{i}{v}\mathbf{w}(x)\right) \quad \text{with } \mathbf{w}(x) = w^a(x)\,\tau^a, \quad a = 1,2,3, \tag{2.69}$$

where v is the parameter introduced above (2.21), fixed by normalizing the kinetic energy terms for the scalar fields w^a. These scalars are the Goldstone bosons associated to the spontaneous breaking of the electroweak symmetry.

The parameterization (2.69) is not unique. One may choose as well

$$\Sigma(x) = \frac{1}{\sqrt{1 + \mathbf{w}^2/v^2}} \left(\mathbf{1} - i\mathbf{w}/v \right) \tag{2.70}$$

or any other representation of a unitary matrix. Any two parameterizations can be related by a nonlinear field redefinition; such a redefinition does not affect the S matrix [23]. However, the Feynman rules and Greens functions for any two parameterizations are different.

In the exponential parameterization (2.69), the leading Goldstone interactions are

$$\Sigma = \left(1 - \frac{1}{2v^2} w^a w^a \pm \cdots \right) \mathbf{1} - \frac{i}{v} \left(1 - \frac{1}{6v^2} w^b w^b \pm \cdots \right) . w^a \tau^a \tag{2.71}$$

The vector field V_μ (2.54) has the expansion (omitting the gauge couplings)

$$V_\mu = \left(\frac{i}{v} \delta^{ab} + \frac{1}{v^2} \epsilon^{abc} w^c \pm \cdots \right) \partial_\mu w^a \tau^b. \tag{2.72}$$

Allowing for fluctuation of $\mathrm{tr}\left[\Sigma^\dagger \Sigma \right]$, the Higgs boson can be introduced by the replacement

$$\Sigma \rightarrow \left(1 + \frac{1}{v} H \right) \Sigma, \tag{2.73}$$

such that

$$\frac{1}{2} \mathrm{tr}\left[\Sigma^\dagger \Sigma \right] = \left(1 + \frac{1}{v} H \right)^2. \tag{2.74}$$

If the constraint (2.68) is imposed, the potential $\mathcal{L}_\Sigma$ (2.58) reduces to a constant. Otherwise, it produces a potential term for the H field:

$$\mathcal{L}_\Sigma = -\frac{\mu^2 v^2}{2} \left(1 + \frac{1}{v} H \right)^2 + \frac{\lambda v^4}{4} \left(1 + \frac{1}{v} H \right)^4 + \cdots . \tag{2.75}$$

To enforce the normalization condition (2.60), one has to require that the minimum of $\mathcal{L}_\Sigma$ occurs at $H = 0$, which implies

$$-\mu^2 = \lambda v^2 + \cdots . \tag{2.76}$$

Again, field redefinitions do not affect the S matrix. One such redefinition is the replacement

$$\left(1 + \frac{1}{v} H \right) \Sigma \rightarrow \left(1 + \frac{1}{v} H \right) - \frac{i}{v} \mathbf{w} \tag{2.77}$$

which parameterizes the same set of matrices, but now in a linear form. While in the parameterization (2.73) the H field is an electroweak singlet (with a nonlinear transformation law for the **w** fields), in the linear parameterization (2.77) the four scalar degrees of freedom transform as a complex electroweak doublet ϕ,

$$\Sigma = 1 + \frac{1}{v}\left(\tilde{\phi}\ \phi\right),\tag{2.78}$$

where

$$\tilde{\phi} = \begin{pmatrix} H - iw^3 \\ -i\sqrt{2}\,w^- \end{pmatrix} \quad \text{and} \quad \phi = \begin{pmatrix} -i\sqrt{2}\,w^+ \\ H + iw^3 \end{pmatrix}.\tag{2.79}$$

The choice of parameterization of the Σ field is arbitrary. Nevertheless, it may be more appropriate to choose either a nonlinear or a linear parameterization. In either case, some operator coefficients may turn out to be unnaturally large or small, which could appear natural in the context of the other parameterization. In particular, the approximate custodial symmetry of the electroweak interactions (Section 2.3.3) is a natural property of the linear representation. This gives a hint that the Higgs boson may indeed be an important part of the spectrum.

2.3 The Chiral Lagrangian

Collecting the results of the previous sections, we have constructed an effective Lagrangian of electroweak interactions, which is valid in the energy range where W and Z bosons are dynamical degrees of freedom:

$$\begin{aligned}
\mathcal{L} = &-\frac{v^2}{4}\,\mathrm{tr}\,[V_\mu V^\mu] - \beta'\frac{v^2}{8}\,\mathrm{tr}\,[TV_\mu]\,\mathrm{tr}\,[TV^\mu] \\
&- (\bar{Q}_L \Sigma M_Q Q_R + \bar{L}_L \Sigma M_L L_R + \text{h.c.}) \\
&- \bar{L}_L^c \Sigma^* M_{N_L}\frac{1+\tau^3}{2}\Sigma L_L - \bar{L}_R^c M_{N_R}\frac{1+\tau^3}{2}L_R \\
&+ \bar{Q}_L i\slashed{D}Q_L + \bar{Q}_R i\slashed{D}Q_R + \bar{L}_L i\slashed{D}L_L + \bar{L}_R i\slashed{D}L_R \\
&- \tfrac{1}{2}\,\mathrm{tr}\,[\mathbf{W}_{\mu\nu}\mathbf{W}^{\mu\nu}] - \tfrac{1}{2}\,\mathrm{tr}\,[\mathbf{B}_{\mu\nu}\mathbf{B}^{\mu\nu}] \\
&- \frac{\mu^2 v^2}{4}\,\mathrm{tr}\,[\Sigma^\dagger \Sigma] + \frac{\lambda v^4}{16}\,\mathrm{tr}\,[\Sigma^\dagger \Sigma]^2 + \cdots .
\end{aligned}\tag{2.80}$$

The Higgs potential in the last line is needed only if fluctuations of the Σ field are allowed, i.e., a Higgs boson is present.

This expression is referred to as the *chiral Lagrangian*. The term is often associated with the specific version containing no Higgs boson, i.e., the nonlinear realization of the Σ field. However, since left-handed and right-handed fermions transform differently under the electroweak symmetry group, any Lagrangian describing electroweak interactions is chiral. We will use the term in this generic sense, without the reference to a particular Higgs representation.

The effective Lagrangian (2.80) is not yet complete. Below we will introduce the missing pieces. Once those extra terms are included, we have a completely general description of the known particles and their interactions in the presence of the electroweak symmetries, useful in the whole energy range where perturbation theory is applicable. Any other parameterization of electroweak interactions can be related to (2.80) by a suitable field redefinition, which has no effect on the S matrix.

2.3.1 Gauge Fixing and Ghosts

A unitary quantum field theory for self-interacting vector bosons can be defined only if there is a corresponding gauge symmetry. In the case of electroweak interactions, this is local $SU(2)_L \times U(1)_Y$ invariance. For a perturbative formulation, this local symmetry must be broken by hand, introducing a gauge-fixing term. When constructing the effective action at higher orders, maintaining unitarity in scattering processes, one has to replace local gauge invariance by a global symmetry, BRS invariance [24]. For the physical fields, this symmetry coincides with gauge invariance. To make the gauge-fixing term BRS invariant one has to introduce additional ghost fields and a corresponding interaction term. This setup is necessary for correctly defining interactions at short distances (absorbing UV divergences), but it does not interfere with spontaneous symmetry breaking which is a long-distance phenomenon.

The gauge-fixing term introduced by 't Hooft [3] is a quadratic form,

$$\mathcal{L}_{\text{gf}} = -\tfrac{1}{2}(F^a F^a + F^0 F^0), \qquad (2.81)$$

which in our notation is given by

$$F^a = \frac{1}{\sqrt{\xi^a}} \partial^\mu W_\mu^a + \frac{iv}{2} \sqrt{\xi^a}\, M_W \operatorname{tr}\left[\Sigma \tau^a\right], \qquad (a = 1, 2, 3) \qquad (2.82)$$

$$F^0 = \frac{1}{\sqrt{\xi^0}} \partial^\mu B_\mu - \frac{iv}{2} s_w \sqrt{\xi^0}\, M_Z \operatorname{tr}\left[\Sigma \tau^3\right]. \qquad (2.83)$$

with four free parameters ξ^a and ξ^0. This form is not unique, but it has two important properties. It is linear in the vector fields which restricts the form of necessary counterterms in an actual calculation, and it cancels all bilinear $W\Sigma$ and $Z\Sigma$ mixing terms which otherwise would require a re-diagonalization of external fields in S matrix elements.

The gauge-fixing term determines, to leading order, the form of the ghost interaction term. One introduces fermionic ghost fields u^a, u^0 and antighost fields $\bar{u}^a, \bar{u}^0$ for each component of the vector fields. To derive their interaction Lagrangian, the Fadeev-Popov term

$$\mathcal{L}_{\text{FP}} = \sqrt{\xi^a}\, \bar{u}^a (\delta^b F^a) u^b + \sqrt{\xi^0}\, \bar{u}^0 (\delta^0 F^0) u^0, \qquad (2.84)$$

we need the gauge variations with respect to $SU(2)_L$

$$\delta^b \Sigma = -ig\frac{\tau^b}{2}\Sigma, \qquad \delta^b W_\mu^a = \delta^{ab}\partial_\mu + g\epsilon^{abc}W_\mu^c, \qquad \delta^b B_\mu = 0, \qquad (2.85)$$

and $U(1)_Y$,

$$\delta^0 \Sigma = ig'\Sigma\frac{\tau^3}{2}, \qquad \delta^0 W_\mu^a = 0, \qquad \delta^0 B_\mu = \partial_\mu. \qquad (2.86)$$

The result is

$$\mathcal{L}_{\mathrm{FP}} = \sum_a \bar{u}^a \left(\Box + \xi^a M_W^2 \tfrac{1}{2}\operatorname{tr}[\Sigma]\right) u^a$$
$$+ \sum_{abc} \epsilon^{abc}\bar{u}^a \left(\partial^\mu W_\mu^c + i\xi^a M_W^2 \tfrac{1}{2}\operatorname{tr}[\Sigma\tau^c]\right) u^b \qquad (2.87)$$
$$+ \bar{u}^0 \left(\Box + \xi^0 s_w^2 M_Z^2 \tfrac{1}{2}\operatorname{tr}[\Sigma]\right) u^0.$$

2.3.2 Anomalous Couplings

The guideline for constructing the effective Lagrangian (2.80) has been to start with the Fermi theory Lagrangian and add the minimal set of fields that make the weak-interaction symmetries manifest. This requires the addition of kinetic terms for the new fields and the inclusion of Σ factors in the boson and fermion mass terms. However, if one does one-loop calculations with (2.80), gauge-fixing, and ghost terms taken into account, one will observe that additional operators are needed as counterterms to make the theory finite at next-to-leading order. This is natural since the Lagrangian (2.80) does not yet contain all possible dimension-four operators consistent with electroweak symmetry. While we have included Σ-dependent terms of dimension two and three, we have not checked for symmetric Σ-dependent terms of higher dimension.

The complete list of CP-invariant[5] dimension-four operators not contained in (2.80) reads [10]:

$$\mathcal{L}_1 = \alpha_1 g g' \operatorname{tr}\left[\Sigma\mathbf{B}_{\mu\nu}\Sigma^\dagger\mathbf{W}^{\mu\nu}\right], \qquad (2.88)$$
$$\mathcal{L}_2 = i\alpha_2 g' \operatorname{tr}\left[\Sigma\mathbf{B}_{\mu\nu}\Sigma^\dagger[V^\mu, V^\nu]\right], \qquad (2.89)$$
$$\mathcal{L}_3 = i\alpha_3 g \operatorname{tr}\left[\mathbf{W}_{\mu\nu}[V^\mu, V^\nu]\right], \qquad (2.90)$$
$$\mathcal{L}_4 = \alpha_4 (\operatorname{tr}[V_\mu V_\nu])^2, \qquad (2.91)$$
$$\mathcal{L}_5 = \alpha_5 (\operatorname{tr}[V_\mu V^\mu])^2, \qquad (2.92)$$
$$\mathcal{L}_6 = \alpha_6 \operatorname{tr}[V_\mu V_\nu] \operatorname{tr}[TV^\mu] \operatorname{tr}[TV^\nu], \qquad (2.93)$$
$$\mathcal{L}_7 = \alpha_7 \operatorname{tr}[V_\mu V^\mu] \operatorname{tr}[TV_\nu] \operatorname{tr}[TV^\nu], \qquad (2.94)$$
$$\mathcal{L}_8 = \tfrac{1}{4}\alpha_8 g^2 (\operatorname{tr}[T\mathbf{W}_{\mu\nu}])^2, \qquad (2.95)$$
$$\mathcal{L}_9 = \tfrac{1}{2}\alpha_9 g \operatorname{tr}[T\mathbf{W}_{\mu\nu}] \operatorname{tr}[T[V^\mu, V^\nu]], \qquad (2.96)$$
$$\mathcal{L}_{10} = \tfrac{1}{2}\alpha_{10} (\operatorname{tr}[TV_\mu] \operatorname{tr}[TV_\nu])^2, \qquad (2.97)$$
$$\mathcal{L}_{11} = \alpha_{11} g\epsilon^{\mu\nu\rho\lambda} \operatorname{tr}[TV_\mu] \operatorname{tr}[V_\nu \mathbf{W}_{\rho\lambda}]. \qquad (2.98)$$

---

[5] There are also operators which parameterize CP violation in the Higgs sector. We will not consider them in this review.

Two of these operators induce terms bilinear in the gauge fields,

$$\mathcal{L}_1 = \alpha_1 g g' \, \mathrm{tr} \left[\mathbf{B}_{\mu\nu} \mathbf{W}^{\mu\nu} \right] + \dots, \tag{2.99}$$

$$\mathcal{L}_8 = \frac{1}{4} \alpha_8 g^2 (\mathrm{tr} \left[\tau^3 \mathbf{W}_{\mu\nu} \right])^2 + \dots, \tag{2.100}$$

where the dots indicate terms involving Goldstone bosons which disappear in unitary gauge. Together with the standard kinetic operators (2.48), these operators determine the physical vector boson fields. Imposing canonical normalization on the kinetic energy of the physical gauge bosons,

$$\mathcal{L}_{\mathrm{kin}} = -\frac{1}{4} A_{\mu\nu} A^{\mu\nu} - \frac{1}{4} Z_{\mu\nu} Z^{\mu\nu} - \frac{1}{2} W^+_{\mu\nu} W^{-\mu\nu}, \tag{2.101}$$

and requiring Z-A-mixing to vanish, we can express the mixing parameters x_Z, y_Z, x_A, y_A (2.29, 2.30) in terms of α_1 and α_8. The result is

$$x_Z = \frac{1}{2} \alpha_1 g g' \frac{s_w}{c_w} - \frac{1}{2} \alpha_8 g^2, \qquad x_A = -\frac{1}{2} \alpha_1 g g' \frac{c_w}{s_w} - \frac{1}{2} \alpha_8 g^2, \tag{2.102}$$

$$y_Z = \frac{1}{2} \alpha_1 g g' \frac{c_w}{s_w}, \qquad y_A = -\frac{1}{2} \alpha_1 g g' \frac{s_w}{c_w}. \tag{2.103}$$

Inserting this into (2.35) and (2.38), we obtain

$$g = \frac{e}{s_w} \left(1 - (\alpha_1 + \alpha_8) \frac{e^2}{2 s_w^2} \right), \tag{2.104}$$

$$g' = \frac{e}{c_w} \left(1 - \alpha_1 \frac{e^2}{2 c_w^2} \right). \tag{2.105}$$

The W and Z masses are determined by (2.20) and (2.33, 2.34)

$$M_W = \frac{ev}{2 s_w} \left(1 - (\alpha_1 + \alpha_8) \frac{e^2}{2 s_w^2} \right), \tag{2.106}$$

$$M_Z = \frac{ev}{2 s_w c_w} \left(1 - \alpha_1 \frac{e^2}{2 c_w^2 s_w^2} + \frac{\beta'}{2} \right), \tag{2.107}$$

such that the ρ parameter defined by the ratio of vector boson masses [25] is given by

$$\rho \equiv \frac{M_W^2}{c_w^2 M_Z^2} = 1 + \alpha_1 \frac{e^2}{c_w^2} - \alpha_8 \frac{e^2}{s_w^2} - \beta', \tag{2.108}$$

which differs from the low-energy ρ_* parameter (2.66). [Note that we define s_w by its low-energy value in neutral-current interactions (2.16)].

In the nonlinear symmetry representation the Lagrangian contains terms of arbitrarily high dimension once the Σ field is expanded. Therefore, the inclusion of the operators (2.88–2.98) is not sufficient to make the theory finite to *all* orders. In each order of perturbation theory new terms are introduced with the dimension of the Σ-dependent terms increased by two.

For instance, the anomalous couplings listed above are generated at one-loop order by insertions of the dimension-two operators tr $[V_\mu V^\mu]$ and tr $[TV_\mu]^2$ and by double insertions of fermion mass operators. Similarly, at one-loop order, insertions of the fermion mass operators like $\bar{Q}_L \Sigma M_Q Q_R$ contracted with the same dimension-two operators will generate dimension-five terms:

$$\mathcal{L}_5 = \bar{Q}_L V_\mu V^\mu \Sigma M'_Q Q_R + \bar{L}_L V_\mu V^\mu \Sigma M'_L L_R + \text{h.c.}, \tag{2.109}$$

$$\mathcal{L}_{5a} = \bar{Q}_L [V_\mu, T] V^\mu \Sigma M'_{Qa} Q_R + \bar{L}_L [V_\mu, T] V^\mu \Sigma M'_{La} L_R + \text{h.c.}, \tag{2.110}$$

$$\mathcal{L}_{5s} = \bar{Q}_L \{V_\mu, T\} V^\mu \Sigma M'_{Qs} Q_R + \bar{L}_L \{V_\mu, T\} V^\mu \Sigma M'_{Ls} L_R + \text{h.c.} \tag{2.111}$$

In principle, the matrices M'_{Qi}, M'_{Li} are independent matrices in flavor space, but as far as these operators are counterterms induced at one-loop order, in a flavor-independent renormalization scheme they are aligned to M_Q and M_L, respectively.

If there is a Higgs boson such that fluctuations of the modulus of Σ are allowed, one can introduce another class of independent operators by multiplying factors of $\frac{1}{2}$ tr $[\Sigma^\dagger \Sigma]$ to any term of the lowest-order Lagrangian. This leads to anomalous Higgs couplings (cf. Section 3.3.1). At dimension six, many additional operators can be constructed, restricted only by (approximate) symmetries one wishes to impose, particularly in the flavor sector [26, 28].

The necessity for higher-dimensional operators does not make the low-energy effective Lagrangian useless. It rather states that one should be prepared for new contributions in the next order of the perturbative series which are at least of the magnitude $1/16\pi^2$ (since they are induced by radiative corrections) with the operator dimension increased by two [11]. The two extra powers of fields or derivatives are compensated by two powers of $1/v$, the expansion parameter of Σ (2.69, 2.70). In matrix elements, the loop expansion is thus equivalent to a power series in terms of

$$\frac{E^2}{\Lambda^2} = \frac{E^2}{(4\pi v)^2}, \tag{2.112}$$

where E is any linear combination of energies, masses, and momenta assigned to the external particles. This sets the scale where perturbation theory breaks down:

$$\Lambda = 4\pi v \approx 3\,\text{TeV}. \tag{2.113}$$

Of course, there may be larger contributions of the operators (2.88–2.98) than predicted by the loop expansion. Since the effect of anomalous couplings increases with energy, generically this leads to a lower scale $\Lambda' < \Lambda$ where perturbation theory breaks down. One would attribute this to independent (tree-level) dynamics not expected from weak interactions. Conversely, assuming all anomalous couplings to vanish in the tree approximation is equivalent to the assumption that the effective Lagrangian is applicable without modification up to the highest possible energy scale.

If the breakdown of the chiral Lagrangian in its minimal form is caused by *additional* degrees of freedom appearing at intermediate scales, the framework is easily extended to include them in the effective Lagrangian. In a particular case (scalar resonances with appropriate quantum numbers, i.e., Higgs bosons), this can soften the ultraviolet behavior, leading to a higher breakdown scale of the new extended effective theory. For special values of the scalar couplings the model allows for a linear symmetry representation. In that case, all loop divergences which call for higher-dimensional counterterms cancel exactly, and the coefficients of anomalous terms could be consistently set to zero without renormalization problems.

This is neither necessary, nor is it even expected to happen. As we will discuss later, essentially all models of electroweak symmetry breaking predict additional matter and interactions at accessible energy scales which cause deviations from the minimal Higgs couplings.

If no Higgs-like state (i.e., a scalar with couplings to massive vector bosons) appears in the spectrum, there is no way to lift the breakdown scale of the chiral Lagrangian. This means that near $\Lambda \approx 3\,\text{TeV}$ the known degrees of freedom can no longer be regarded as elementary in the context of perturbative quantum field theory. They could turn out to be composite states, and it is conceivable that even more fundamental conditions start to break down at this scale, such as the point-like behavior of particles, dimensionality of spacetime, or four-dimensional Lorentz invariance.

2.3.3 Custodial Symmetry

As we have mentioned in Section 2.1.5, the ρ_* parameter of neutral-current interactions is very close to one (radiative corrections properly taken into account), so due to (2.66) we should postulate that the coefficient β' of the operator

$$-\beta' \frac{v^2}{8} \operatorname{tr}[TV_\mu] \operatorname{tr}[TV^\mu] \qquad (2.114)$$

in the chiral Lagrangian (2.80) vanishes to leading order in perturbation theory.

This can be attributed to an approximate symmetry [25]. If we take the global symmetry of the Lagrangian not to be $SU(2)_L \times U(1)_Y$ but $SU(2)_L \times SU(2)_R$ where, as far as the Higgs sector is concerned, $U(1)_Y \subset SU(2)_R$, and impose the transformation law on Σ,

$$\Sigma \to U \Sigma V^\dagger \qquad (2.115)$$

with $U \in SU(2)_L$ and $V \in SU(2)_R$, then the term (2.114) is forbidden. The vector bosons are assumed to be singlets under $SU(2)_R$.

It appears natural to take the right-handed fermions to be doublets under $SU(2)_R$, just as the left-handed fermions are doublets under $SU(2)_L$. Unfortunately, for the fermions this is not a symmetry. If it were, up- and down quark masses would be equal and the CKM mixing matrix would be trivial. Furthermore, if this symmetry were exact the Majorana mass terms of the neutrinos would have to vanish, forcing them to be Dirac particles and leaving behind an extra unbroken global $U(1)$ symmetry at low energies. The nature of neutrino masses cannot currently be decided from data, but the nontriviality of the quark mixing matrix and the up-down quark mass splittings are experimental facts. Clearly, the strongest violation of this symmetry comes from the large top quark mass.

Moreover, weak interactions themselves break this symmetry. The B vector field, in our notation, couples to Σ by a τ^3 matrix factor, which is not consistent with a right-handed $SU(2)_R$ symmetry. Nevertheless, this breaking is proportional to s_w^2 which is not a large parameter, and the fermion couplings affect vector and scalar interactions at the loop level only. Looking at bosonic interactions, $SU(2)_R$ invariance is not a bad approximation.

The limit of exact $SU(2)_R$ symmetry in the Lagrangian is realized by setting up- and down-Yukawa couplings equal and

$$e \to 0 \quad \text{and} \quad s_w \to 0, \quad e/s_w \text{ fixed.} \tag{2.116}$$

In this limit, the nonvanishing asymptotic expectation value of Σ induces an enlarged symmetry-breaking pattern

$$SU(2)_L \times SU(2)_R \to SU(2)_C, \tag{2.117}$$

where $SU(2)_C$ is the diagonal symmetry under which Σ transforms as

$$\Sigma \to C \Sigma C^\dagger \tag{2.118}$$

with a matrix $C \in SU(2)_C$, such that the expectation value $\langle \Sigma \rangle = 1$ is invariant.

Under $SU(2)_C$ transformations the W and Z vector fields make up a mass-degenerate triplet. Indeed, with

$$1/c_w^2 \approx 1 + s_w^2 \tag{2.119}$$

we read off from (2.108) that in the symmetric limit $s_w = 0$ one has $M_W = M_Z$ to leading order, and the relation $\rho = 1$ is exact. This feature is responsible for $SU(2)_C$ being called a *custodial symmetry* since it keeps deviations from $\rho = 1$ small. The tree-level correction to the mass degeneracy $M_W = M_Z$ is proportional to s_w^2.

At the loop level there will be additional small corrections proportional to $g^2 s_w^2 / 16\pi^2$ and proportional to the fermion mass splittings squared, in particular $(m_t^2 - m_b^2)/16\pi^2$. Furthermore, in (2.108) there are additional corrections proportional to α_1, α_8, and β', where in the symmetric limit the contribution of α_1 vanishes. The coefficients α_8 and β' multiply operators (2.88, 2.95) which induce explicit custodial $SU(2)_C$ violation due to the fact that their algebraic expressions contain factors of T (2.54).

The explicit breaking of $SU(2)_C$ in the Lagrangian makes the operator (2.114) a counterterm necessary to remove divergences at next-to-leading order (in a nonlinear representation). Thus, its coefficient cannot be exactly zero but is expected to be of the order $1/16\pi^2$. Concerning the operators (2.88–2.98), if (2.114) is absent in the leading-order effective action, only the operators that do not contain explicit T factors are needed as one-loop counterterms,

$$\mathcal{L}_1 = \alpha_1 g g' \operatorname{tr} \left[\Sigma \mathbf{B}_{\mu\nu} \Sigma^\dagger \mathbf{W}^{\mu\nu} \right], \tag{2.120}$$

$$\mathcal{L}_2 = i\alpha_2 g' \operatorname{tr} \left[\Sigma \mathbf{B}_{\mu\nu} \Sigma^\dagger [V^\mu, V^\nu] \right], \tag{2.121}$$

$$\mathcal{L}_3 = i\alpha_3 g \operatorname{tr} \left[\mathbf{W}_{\mu\nu} [V^\mu, V^\nu] \right], \tag{2.122}$$

$$\mathcal{L}_4 = \alpha_4 (\operatorname{tr} [V_\mu V_\nu])^2, \tag{2.123}$$

$$\mathcal{L}_5 = \alpha_5 (\operatorname{tr} [V_\mu V^\mu])^2, \tag{2.124}$$

while the others arise first at two-loop order, accompanied by $SU(2)_C$-symmetric operators of dimension eight. This constraint drastically reduces the number of independent coefficients, and it leads to predictions that should be testable in experiment. However, the fact that custodial symmetry violation is apparently small at leading order may be accidental, and the $SU(2)_C$-violating operators of dimension six may have larger coefficients than required by renormalization consistency.

Similar arguments hold for fermionic operators. Dimension-five counterterms arise from Feynman diagrams where the fields parameterizing Σ are contracted between terms from $\mathcal{L}_3$ (fermion mass terms) and $\mathcal{L}_{2(W)}$ (vector boson masses generated by the Σ kinetic energy operator). Again, the absence of the operator (2.114) at leading order eliminates one-loop counterterms containing explicit T factors. The remaining term has the form

$$\mathcal{L}_5 = \bar{Q}_L V_\mu V^\mu \Sigma M'_Q Q_R + \bar{L}_L V_\mu V^\mu \Sigma M'_L L_R + \text{h.c.} \tag{2.125}$$

If other fermionic operators are present at this order, they originate from extra $SU(2)_C$ violation not encountered at leading order.

2.3.4 The Chiral Lagrangian Revisited

The approximate custodial $SU(2)_C$ and the exact electromagnetic gauge invariance are symmetries that survive at low energies, while the full electroweak symmetry $SU(2)_L \times U(1)_Y$ is broken. To make the low-energy structure more explicit, one can rewrite the chiral Lagrangian (2.80) in a particular way [9]. One combines left- and right-handed quark and lepton fields in Dirac multiplets, such that coupling additional massive fields associated with electroweak symmetry breaking is straightforward.

We introduce the square root of the Σ field,

$$\Sigma = \xi\xi \quad \text{and} \quad \Sigma^\dagger = \xi^\dagger\xi^\dagger, \tag{2.126}$$

which again is a unitary 2×2 matrix field,

$$\xi^\dagger \xi = \xi \xi^\dagger = 1, \tag{2.127}$$

if there is no Higgs boson. In the exponential representation for Σ, ξ is also an exponential

$$\xi = \exp\left(-\frac{i}{2v}\mathbf{w}\right). \tag{2.128}$$

If a Higgs boson is present, one can define ξ by taking the square root of the Higgs field as well,

$$\xi \rightarrow \sqrt{1 + \frac{1}{v}H}\,\xi = \left(1 + \frac{1}{2v}H + \ldots\right)\xi, \tag{2.129}$$

or one may write

$$\xi \rightarrow \left(1 + \frac{1}{2v}H\right)\xi, \tag{2.130}$$

such that Σ is proportional to $(1 + \frac{1}{2v}H)^2$. In the linear representation this introduces a nonlinearity, so coupling terms involving ξ explicitly can spoil renormalizability.

We proceed to state the transformation properties of ξ under $SU(2)_L \times SU(2)_R$. The transformation law of Σ,

$$\Sigma(x) \rightarrow U(x)\,\Sigma(x)\,V(x)^\dagger, \tag{2.131}$$

can be satisfied if we have

$$\xi(x) \rightarrow U(x)\,\xi(x)\,C(x)^\dagger = C(x)\,\xi(x)\,V(x)^\dagger, \tag{2.132}$$

where $C(x)$ has yet to be determined. If we know $U(x)$ and $V(x)$, for any given field configuration $C(x)$ can be computed from (2.132). It is the matrix which solves

$$C\xi V^\dagger C\xi^\dagger U^\dagger = 1. \tag{2.133}$$

Clearly, $C(x)$ is a function not only of the group elements $U(x)$ and $V(x)$, but also of $\xi(x)$.

If we restrict ourselves to *global* $SU(2)_L \times SU(2)_R$ transformations, we have constant matrices U and V, but (2.133) shows that C is position-dependent nevertheless. Only if $U = V$,

$$C = U = V \tag{2.134}$$

is always a solution. In this case all three matrices are position-independent. In the general case $C(x)$ depends on position via the ξ fields even though the original chiral transformation (U, V) is global. As a consequence, the effective Lagrangian is invariant under position-dependent $SU(2)_C$ transformations, i.e., it looks like a gauge theory for the $SU(2)_C$ symmetry. Nevertheless, this symmetry does not impose the constraints of the full *local* $SU(2)_L \times U(1)_Y$ invariance which is a stronger restriction.

As another special case, for local transformations with $U(x) = V(x) \in U(1)$, $C(x)$ describes electromagnetic gauge transformations,

$$C(x) = U(x) = V(x). \tag{2.135}$$

In the general case, $C(x)$ transformations depend on the Goldstone fields $\mathbf{w}(x)$, so the electroweak symmetry group is represented in a nonlinear way. Again, in unitary gauge we set $\xi = 1$ which is consistent with both $SU(2)_C$ and $U(1)_{\mathrm{em}}$ symmetry.

Four-component Dirac spinors are built up from left- and right-handed fields by

$$Q = \begin{pmatrix} \xi Q_R \\ \xi^\dagger Q_L \end{pmatrix} \quad \text{and} \quad L = \begin{pmatrix} \xi L_R \\ \xi^\dagger L_L \end{pmatrix}, \tag{2.136}$$

such that there is a common transformation law

$$Q \to CQ \quad \text{and} \quad L \to CL. \tag{2.137}$$

The chiral fields are projected out by the γ_5 matrix,

$$Q_L = \xi \tfrac{1}{2}(1 - \gamma_5)Q, \quad Q_R = \xi^\dagger \tfrac{1}{2}(1 + \gamma_5)Q, \tag{2.138}$$

and analogously for L_L, L_R. Other fields in the Lagrangian are replaced as follows:

$$\mathbf{W}_\mu \Rightarrow \xi \mathbf{W}_\mu \xi^\dagger, \quad V_\mu \Rightarrow \xi V_\mu \xi^\dagger, \quad T \Rightarrow \xi T \xi^\dagger, \tag{2.139}$$

to eliminate explicit ξ and $\xi^\dagger$ factors in the Lagrangian and in the anomalous couplings $\mathcal{L}_3 \dots \mathcal{L}_{11}$ (2.88–2.98).

This does not work for the B field. Looking at the operators $\mathcal{L}_1$ and $\mathcal{L}_2$ (2.88, 2.89), we could make the replacement

$$\mathbf{B}_\mu \Rightarrow \xi^\dagger \mathbf{B}_\mu \xi, \tag{2.140}$$

which eliminates ξ and $\xi^\dagger$ factors there as well (note the exchange of ξ and $\xi^\dagger$). However, no transformation law can make this happen in the derivatives involving B_μ (2.39, 2.40) since the couplings proportional to the matrices q and τ^3 break the custodial symmetry:

$$qB_\mu = 2q\tau^3 \mathbf{B}_\mu \Rightarrow 2q\tau^3 \xi^\dagger \mathbf{B}_\mu \xi. \tag{2.141}$$

The ξ factors disappear from the Lagrangian completely only if we neglect this coupling, i.e., the $U(1)_Y$ gauge group. This is the limit $s_w \to 0$, g constant. The same is true for the up-down mass splittings in the fermion sector.

In the limit of exact $SU(2)_C$ symmetry, in terms of the new fields the quark interactions assume the form

$$\bar{Q}_L \mathrm{i} \slashed{D} Q_L + \bar{Q}_R \mathrm{i} \slashed{D} Q_R = \bar{Q}(\mathrm{i}\slashed{\partial} + \slashed{V})Q - \bar{Q}\slashed{A}\gamma_5 Q - g\bar{Q}\slashed{\mathbf{W}}Q, \tag{2.142}$$

analogously for the leptons. Here, we have introduced two abbreviations which serve as composite "vector" and "axial vector" fields

$$\mathcal{V}_\mu = \frac{i}{2}(\xi^\dagger \partial_\mu \xi + \xi \partial_\mu \xi^\dagger), \tag{2.143}$$

$$\mathcal{A}_\mu = \frac{i}{2}(\xi^\dagger \partial_\mu \xi - \xi \partial_\mu \xi^\dagger). \tag{2.144}$$

Under global $SU(2)_L \times SU(2)_R$ transformations (for which the matrix C in general is position-dependent), the $\mathbf{W}$ vector field transforms homogeneously, and the replacement (2.139) makes its transformation law

$$\mathbf{W}_\mu \to C\mathbf{W}_\mu C^\dagger. \tag{2.145}$$

Thus, $\mathbf{W}_\mu$ is *not* the gauge field of $SU(2)_C$. As a consequence, the third term on the right-hand side of (2.142) is invariant on its own.

Noting that the left-hand side of (2.142) is invariant, and that the transformation law of the quark kinetic energy is inhomogeneous due to the position dependence of C, one can deduce the transformation laws of the composite vector and axial vector fields

$$\mathcal{V}_\mu \to C\mathcal{V}_\mu C^\dagger - i(\partial_\mu C)C^\dagger \tag{2.146}$$

$$\mathcal{A}_\mu \to C\mathcal{A}_\mu C^\dagger. \tag{2.147}$$

Apparently, for the $SU(2)_C$ symmetry, the $\mathcal{V}$ field plays the role of a gauge boson. Due to this fact, the covariant derivatives

$$\mathcal{D}_\mu Q \equiv \partial_\mu Q - i\mathcal{V}_\mu Q \quad \text{and} \quad \mathcal{D}_\mu \mathbf{W}_\nu \equiv \partial_\mu \mathbf{W}_\nu - i\mathcal{V}_\mu \mathbf{W}_\nu + i\mathbf{W}_\nu \mathcal{V}_\mu \tag{2.148}$$

transform covariantly for position-dependent transformations $C(x)$.

An invariant kinetic-energy term for the W_μ field can accordingly be written as

$$-\tfrac{1}{2}\operatorname{tr}[\mathbf{W}_{\mu\nu}\mathbf{W}^{\mu\nu}] \quad \text{with} \quad \mathbf{W}_{\mu\nu} = \mathcal{D}_\mu \mathbf{W}_\nu - \mathcal{D}_\nu \mathbf{W}_\mu + ig[\mathbf{W}_\mu, \mathbf{W}_\nu], \tag{2.149}$$

but independent triple and quartic gauge boson couplings are also consistent with the symmetry, e.g.,

$$\operatorname{tr}[(\mathcal{D}_\mu \mathbf{W}_\nu)\mathbf{W}^\mu \mathbf{W}^\nu], \quad \operatorname{tr}[\mathbf{W}_\mu \mathbf{W}_\nu \mathbf{W}^\mu \mathbf{W}^\nu], \quad \text{etc.} \tag{2.150}$$

Furthermore, the homogeneous transformation laws of $\mathbf{W}$ (2.145) and $\mathcal{A}$ (2.147) allow for writing "mass terms":

$$\mathcal{L}_\Sigma = 2v^2 \operatorname{tr}[\mathcal{A}_\mu \mathcal{A}^\mu] + \tfrac{1}{2}M_W^2 \operatorname{tr}[\mathbf{W}_\mu \mathbf{W}^\mu]. \tag{2.151}$$

Expanding ξ in terms of Goldstone fields, one observes that the first term contains the Goldstone boson kinetic energy while the second term is responsible for the W mass. We can therefore identify this operator with $\mathcal{L}_\Sigma$ (2.58).

Again, the symmetry does not impose a relation between the two operator coefficients v^2 and M_W^2 and the coupling g. Such relations follow from imposing a *local* $SU(2)_L$ symmetry, while the manifest symmetry in the present formulation is just spontaneously broken *global* $SU(2)_L \times SU(2)_R$ invariance, translated into local $SU(2)_C$ invariance. Incidentally, the arbitrariness in writing bilinear, cubic, and quartic couplings for the W_μ field (and for the B_μ field if it is re-introduced) makes the operators $\mathcal{L}_{1\ldots11}$ (2.88–2.98) redundant in this formulation.

Apparently, this alternative version of the chiral Lagrangian is not well suited for describing particles with well-defined transformation properties under local $SU(2)_L \times U(1)_Y$. Its principal application is the description of states which are connected to the symmetry-*breaking* sector. If this sector is strongly interacting, there should be states with definite $SU(2)_C$ quantum numbers, but no assignment to complete $SU(2)_L \times U(1)_Y$ (or $SU(2)_L \times SU(2)_R$) multiplets. The interactions should show a trace of global $SU(2)_L \times SU(2)_R$ invariance, represented by local $SU(2)_C$ invariance with the composite vector field $\mathcal{V}_\mu$ playing the role of the gauge bosons. In low-energy QCD, this happens for hadronic resonances, mesons, and baryons, and it could apply to the dynamics of electroweak symmetry breaking as well.

For instance, a heavy Dirac field Ψ (*technibaryon*) which does not live in a $SU(2)_L \times U(1)_Y$ representation but in a $SU(2)_C$ doublet, could mix with ordinary fermions via the bilinear coupling

$$\bar{Q}M\Psi + \text{h.c.} = \bar{Q}_L \xi M_L \tfrac{1}{2}(1 + \gamma_5)\Psi + \bar{Q}_R \xi^\dagger M_R \tfrac{1}{2}(1 - \gamma_5)\Psi + \text{h.c.} \quad (2.152)$$

Its electroweak couplings can be written just as for the quarks and leptons by

$$\bar{\Psi}(i\partial\!\!\!/ + \mathcal{V}\!\!\!\!/)\Psi + g_A \bar{\Psi}\mathcal{A}\!\!\!/\Psi + g_W \bar{\Psi}\mathbf{W}\!\!\!\!/\Psi \quad (2.153)$$

with unknown values for the axial-vector and the weak couplings g_A and g_W. In general, one would also expect anomalous gauge couplings such as

$$\bar{\Psi}\xi^\dagger \mathcal{V}\!\!\!\!/\xi\Psi, \quad \bar{\Psi}\xi^\dagger \mathbf{W}_{\mu\nu}\sigma^{\mu\nu}\xi\Psi. \quad (2.154)$$

If $SU(2)_C$ is broken, one should add terms involving T. In case there are no $SU(2)_C$ partners, one can add dummy ones and extract the physical states using $\tfrac{1}{2}(1 \pm T)$ projectors.

2.3.5 The Linear Representation

A $SU(2)_C$ singlet scalar is a very special resonance state. This serves as a Higgs boson if its couplings (at least, approximately) correspond to $\Sigma^\dagger \Sigma$ fluctuations. As mentioned at the end of Section 2.2.4, in this case a linear representation (2.77) of Σ may be more appropriate to describe the phenomenology. If a linear representation is adopted, the power-counting in the perturbative series can be made more precise, since

$$\Sigma = \frac{1}{v}\left((v + H) - i\mathbf{w}\right) \tag{2.155}$$

contains all fields to first power only. In this situation it is appropriate to assign dimension one to Σ and make the $1/v$ prefactor explicit. The shorthands V_μ and T (2.54) should be expanded, of course.

With the modified power counting, in the original chiral Lagrangian (2.80) two operators get a dimension higher than four and thus become formally irrelevant. First, the left-handed neutrino mass operator gets dimension five, such that lepton number conservation becomes natural at low energies where no right-handed neutrinos are present. Second, the custodial-symmetry violating operator

$$-\beta' \frac{v^2}{8} \operatorname{tr}\left[TV_\mu\right] \operatorname{tr}\left[TV^\mu\right] = -\beta' \frac{v^2}{8} \operatorname{tr}\left[\tau^3 (D_\mu \Sigma)^\dagger \Sigma\right] \operatorname{tr}\left[(D_\mu \Sigma)\Sigma^\dagger \tau^3\right] \tag{2.156}$$

turns out to have dimension six. Applying the same power-counting to the anomalous operators $\mathcal{L}_1 \ldots \mathcal{L}_{11}$, the first three become of dimension six while the remaining operators get dimension eight. Similar statements hold for the fermionic operators. Renormalization no longer requires any of those to be present in the effective Lagrangian [3].

In other words, in a model with a dynamic Higgs boson custodial symmetry is automatically manifest in the renormalizable part of the Lagrangian (up to the perturbations due to B_μ couplings and the fermionic up-down mass splittings). The situation changes below the Higgs mass scale where the power-counting of the nonlinear representation applies. As a consequence, one should expect that $SU(2)_C$-violating contributions originating from physics above the Higgs mass scale are naturally suppressed by a factor of v^2/Λ^2 where Λ is a typical scale for this kind of new physics. If $\Lambda \approx 4\pi v$ (the no-Higgs case, cf. Section 2.3.2), this is equivalent to a loop factor $1/16\pi^2$. In the effective theory below the Higgs mass scale, $SU(2)_C$ violation is unsuppressed, and divergences appear that are cut off by the Higgs boson mass. This applies to the Higgs boson itself. If it is heavy enough, its absence in the low-energy effective field theory introduces one-loop corrections proportional to $\log m_H$ to the higher-dimensional operators in the low-energy effective theory, including $SU(2)_C$-violating ones.

If $SU(2)_C$ violation in the Higgs sector is small (as suggested by $\rho \approx 1$), one can draw the conclusion that any extra $SU(2)_C$-violating physics should pertain to scales considerably higher than m_H. If the Higgs boson is light enough, the suppression is natural. In the opposite case these new effects — which include top mass generation which strongly violates $SU(2)_C$ — must be somewhat decoupled from the Higgs sector responsible for electroweak symmetry breaking. Turning the argument around, if one assumes that electroweak symmetry breaking and top mass generation are closely entangled, one can derive a rough upper bound for the Higgs mass of around 500 GeV [29]. These assumptions do not necessarily hold, but it has turned out to be difficult to find realistic models which evade this limit.

2.4 The Standard Model and Beyond

In the linear representation where a Higgs boson is included in the parameterization of the Σ field, the chiral Lagrangian (2.80) reduces to the familiar Standard Model (SM) of electroweak interactions. Compared to the generic chiral Lagrangian, this model has only one additional degree of freedom, the Higgs boson, with one additional parameter, the Higgs mass. It is renormalizable, so no higher-dimensional counterterms are needed, and all observables can be calculated in perturbation theory to any desired order if some measurements are selected which fix the values of the free parameters. Furthermore, the predictions can be extrapolated to high energies, unless there is another threshold where new degrees of freedom must be incorporated.

If this breakdown scale of the SM is within reach, anomalous couplings must be included in the effective Lagrangian as in the nonlinear case [26]. Once all parameters of the renormalizable part are known with sufficient precision, additional measurements should reveal nonzero values for those. Below, we will present some arguments that this may indeed be the case.

2.4.1 Limitations of the Minimal Model

The allowed scale range for the SM is limited by constraints that are not obvious from the Lagrangian structure. We have already introduced the Higgs potential $\mathcal{L}_\Sigma$ (2.58) as part of the chiral Lagrangian (2.80), now identified with the SM Lagrangian. It has one new parameter which can be identified either with the Higgs mass or with the Higgs self-coupling λ. Otherwise, the model includes as free parameters the gauge couplings g and g', the fermion masses which are now re-interpreted as Yukawa couplings $g_f = m_f/v$, and finally the CKM matrix elements. There is one parameter v which has the dimension of mass. These parameters undergo logarithmic renormalization when the scale of measurement is changed.

Inserting the actual values of the parameters as they are extracted from data, a global solution of the renormalization group equations does not result in dramatic effects in any running parameter but λ, which is proportional to the Higgs mass. When the scale is increased, if its low-energy initial value is too small it is driven towards zero, while it becomes infinite at some scale if its initial value is too large. The marginal case occurs at about 130 GeV [30].

If the actual Higgs mass deviates from this marginal value, the running of λ thus sets a limit to the energy range where the SM can be valid as an effective theory. With the current experimental constraints on the Higgs mass [6], this limit is not a very strong one. The breakdown scale for the SM is allowed to be considerably higher than the reach of any conceivable scattering experiment; it could be as large as the Planck scale. If nonzero anomalous couplings are present, they would point more directly to the breakdown of the SM.

A much stronger limitation is enforced if we take seriously the hierarchy problem discussed in the Introduction. The inherent scale of the chiral Lagrangian, either its uppermost breakdown scale $4\pi v = 3\,\text{TeV}$, or, if it reduces to the SM, the Higgs mass m_H, is many orders of magnitude smaller than the Planck mass, the fundamental scale of gravitation. One would like to have a natural explanation for this puzzle.

2.4.2 Solving the Hierarchy Problem

The hierarchy problem [7] could have one of the following solutions:

1. The complete theory which describes quantized geometry (i.e., quantum gravity), electroweak and strong interactions and any new physics at the Planck scale, has a low-energy limit which, for some reason, contains a massless scalar multiplet. Electroweak symmetry prohibits all fermion and vector boson masses below this scale except for the right-handed neutrino masses. In addition, there is a dynamical effect which generates a scale from renormalization group running. If the scalar is coupled to this new dynamics, it may get a negative mass squared proportional to this low scale, resulting in electroweak symmetry breaking.
2. The effective theory below the Planck scale has no scalar states. For the Standard-Model fields, with the exception of right-handed neutrinos, electroweak symmetry prohibits all fermion and vector boson masses. There is a dynamical effect which generates a scale from renormalization group running, producing nonvanishing expectation values for composite operators. If a composite field with Higgs quantum numbers effectively gets a negative mass squared, electroweak symmetry breaking results.
3. At energies above the TeV scale, four-dimensional quantum field theory is no longer a useful description of physics. For instance, there may be more than three space dimensions, spacetime may cease to be a continuum, or particles may be no longer pointlike. This possiblity does not immediately explain the presence of a dimensionless number of 10^{15} in the fundamental laws of Nature, but it removes the need for deriving it from the dynamics of local quantum fields.[6]
4. There is not a single effective theory, but the apparent hierarchy is the result of a complicated pattern of effective theories, each with its own scale range of validity.
5. There is nothing but the Standard Model with a Higgs boson below the Planck scale. The hierarchy is a mere accident.

If the last possibility is not seriously being considered, all conceivable scenarios predict new physics at scales considerably below the Planck scale. This new sector might interact with the visible matter only extremely weakly (e.g.,

---

[6] Recently, extra-dimension models have been developed where the solution of the Einstein equations for the background gravitational field contains a large exponential factor, thus giving a potential explanation for the hierarchy problem in terms of the fundamental space-time geometry [31].

by gravitation). In that case, it would be essentially unobservable. However, one may be luckier, and there could be a wealth of direct signatures for the dynamics generating electroweak symmetry breaking.

From the point of view of a high-energy theory valid near the Planck scale, the Higgs field is a multiplet of massless scalars. The actual masses are so tiny that they should be ignored at that stage. However, neither a manifest gauge symmetry nor a global discrete or continuous symmetry under which the scalars transform linearly can forbid masses for interacting scalars in the effective theory below this scale, if it is a quantum field theory in the usual sense. If a scalar multiplet Φ transforms as $\Phi \to U\Phi$ according to some representation U of a group $\mathcal{G}$, this does always allow a mass term $\Phi^\dagger \Phi$ which is symmetric under group transformations.

Looking more closely at the scenarios described above, within the context of four-dimensional quantum field theory three explanations for this paradox come into mind:

1. The Higgs multiplet Σ is part of a multiplet of Goldstone bosons that results from the spontaneous breaking of an enlarged global symmetry at a scale F significantly above the electroweak scale v.

   Unfortunately, this does not work for the whole hierarchy [32]. Goldstone bosons π^a are degrees of freedom associated to generators T^a of the spontaneously broken symmetry. As we have done it for the nonlinear representation of $SU(2)_L \times U(1)_Y$, we can take them to parameterize a field Ξ,

$$\Xi = \exp\left(-\frac{\mathrm{i}}{F}\pi^a(x)T^a\right), \tag{2.157}$$

   where F is the scale of this symmetry breaking. All Goldstone interactions are expressible in terms of Ξ interactions, and the symmetry of the original theory implies the invariance of the action under field translations

$$\pi \to \pi + \text{const.}, \tag{2.158}$$

   which is a nonlinear symmetry. This does not only forbid Goldstone mass terms, but it prohibits all Goldstone interactions except for derivative interactions, which are irrelevant operators. Since Higgs bosons must have sizable (effective) couplings to other particles and to themselves, if they are among the π^a fields the scale F cannot be arbitrarily high. Stated otherwise, the global symmetry must be *explicitly* broken to allow for nonvanishing Higgs interactions. A large hierarchy cannot be generated by this mechanism.

   Nevertheless, recently models have been constructed where a particular symmetry-breaking pattern induces pseudo-Goldstone masses at two-loop order only, and a nontrivial potential is generated that triggers electroweak symmetry breaking [33]. While such *little-Higgs models* by themselves do not solve the hierarchy problem, they open the interesting possibility that at low energies the mimimal SM with a physical Higgs boson is realized, while dynamical symmetry breaking at a higher scale is responsible for the generation of the electroweak scale, with little effect on the precision data presently available.

2. The Higgs multiplet Σ is a member of a supersymmetric (chiral) doublet [34, 35]. For supersymmetries the above argument does not hold since they do not commute with Lorentz transformations, and therefore their representations involve particles of different spin. If electroweak gauge symmetry forbids a mass term for the partner fermion multiplet of the Higgs bosons, supersymmetry then does so for the scalars. Clearly, supersymmetry is broken at a scale Λ which lies between the Planck scale and the electroweak scale. Below this scale the scalar mass term is no longer forbidden, and therefore it acquires a value proportional to Λ.

In this scenario, the existence of Higgs self-couplings and Higgs-fermion couplings is natural. However, the hierarchy problem is not solved since the scale of supersymmetry breaking has to be explained. Again, one would like to find a mechanism of field condensation in some hidden sector, which must be communicated to the Higgs and the other Standard Model fields by either gauge or gravitational interactions. The parameters which enter the low-energy effective Lagrangians, the soft-breaking parameters, are suppressed by additional factors of Λ/F, where F is a characteristic scale for the *messenger* interactions. Therefore, Λ may be considerably higher than the electroweak scale, and accessing the physics associated with dynamical supersymmetry breaking experimentally will be difficult, if not impossible.

3. Field condensation and scale generation takes place in the visible sector, i.e., there are new fields involved which directly couple to the low-energy degrees of freedom. In this case, there are composite states with Higgs and Goldstone quantum numbers which form at a scale not much above the electroweak scale. At higher energies, there are only massless gauge bosons and chiral fermions for which mass terms are forbidden by the gauge symmetry. This gauge symmetry must be a larger group than the SM group $SU(3)_{\text{QCD}} \times SU(2)_L \times U(1)_Y$. Part of the gauge interactions become strong at TeV scales, dynamically breaking the electroweak gauge symmetry and generating composite scalars, possibly including a Higgs boson.

Many different models of this type exist in the literature [36], where the fermions which feel the new strong interactions may be confined and thus unobservable at low energies [22] (*technicolor*). In some models, they are related to the top quark [37, 38] (*topcolor*) or to neutrinos [39] which could have large Dirac mass terms in spite of their small observed (Majorana) masses (cf. Section 2.1.1) and therefore be strongly coupled to the symmetry breaking sector.

This class of scenarios appears to be most economical, since dynamic effects are taken to break the electroweak symmetry directly. However, realistic models become complicated again, since the presence of Higgs-fermion couplings and their own hierarchy structure must be explained. It turns out that additional broken gauge symmetries have to be introduced for that purpose [40], and multiple scales of gauge-symmetry breaking cannot be avoided.

3

Scattering Amplitudes of Massive Vector Bosons

3.1 Goldstone Boson Scattering

In the chiral-Lagrangian approach described in the preceding chapter (which includes the Standard Model as a special case), electroweak symmetry breaking and fermion mass generation are mediated by a matrix-valued field Σ (Section 2.2.1). If one insists that the symmetries of weak interactions are not just accidental, this field has to be introduced to make them consistent with the existence of masses for fermions and vector bosons, and its transformation properties are fixed. The degrees of freedom that make up Σ are therefore a probe for the mechanism of electroweak symmetry breaking.

As we have seen (Section 2.2.4), the minimal parameterization in terms of three fields is nonlinear, e.g.,

$$\Sigma = \exp\left(-\frac{i}{v}\mathbf{w}\right), \tag{3.1}$$

while the minimal linear parameterization involves four fields,

$$\Sigma = \left(1 + \frac{1}{v}H\right) - \frac{i}{v}\mathbf{w}, \tag{3.2}$$

but the actual structure may be much more complicated. The only property common to all possible parameterizations of Σ is the presence of the scalar triplet field $\mathbf{w}$. This is the triplet of Goldstone bosons associated with spontaneous electroweak symmetry breaking [41]. Whether there are additional fields parameterizing Σ, most notably a Higgs boson, is not known at present. Thus, the only experiment which *guarantees* information about electroweak symmetry breaking is the study of Goldstone boson interactions.

3.1.1 The Goldstone-Boson Equivalence Theorem

The Feynman rules of the chiral Lagrangian with Goldstone bosons included are gauge-dependent. In a spontaneously-broken gauge theory the Goldstone bosons are not part of the physical spectrum. To compute observable S-matrix elements, one should project on the observable vector bosons and fermions. Nevertheless, one can compute S-matrix elements with Goldstone bosons as external states. They are not directly measurable, but they contain important information about Goldstone boson interactions, the window into the physics of electroweak symmetry breaking.

To access these S-matrix elements, one makes use of the Goldstone boson equivalence theorem [42]. The w^a fields in the effective Lagrangian have the same quantum numbers as the divergence of the vector boson fields $\partial \cdot W^a$. Depending on the gauge, there is a mixing term $w^a(\partial \cdot W^a)$ in the effective Lagrangian, which by partial integration can also be read as $W^a \cdot \partial w^a$. Heuristically, this implies that a (unphysical) one-particle state $w^a|0\rangle$ has some overlap with the one-particle state $\partial \cdot W^a|0\rangle$:

$$\langle 0|w^a \partial \cdot W^a|0\rangle \neq 0. \tag{3.3}$$

This can be formally proven by using the BRS transformation properties of the external fields [24].

According to the general S-matrix equivalence theorem [23, 9], two fields which have a nonvanishing overlap can be traded for each other in S matrix calculations. In the present case, an S matrix element with w^a as an asymptotic external field is thus equal to the analogous S-matrix element with the asymptotic external field $\partial \cdot W^a$. Again, this state (the scalar polarization state of the vector boson) is unphysical, but one can look instead at longitudinally polarized W bosons. For those, the polarization vector takes the form

$$\epsilon_L = \frac{1}{M}\left(|\mathbf{p}|; E\frac{\mathbf{p}}{|\mathbf{p}|}\right), \tag{3.4}$$

which has the properties

$$\epsilon_L^2 = -1 \quad \text{and} \quad \epsilon_L \cdot p = 0. \tag{3.5}$$

For $E \gg M$ this polarization vector becomes aligned with the momentum vector:

$$\epsilon_L = \frac{p}{M} + O(\frac{M}{E}). \tag{3.6}$$

Since polarization states are projected onto by contracting matrix elements involving W_μ^a with the appropriate polarization vector, scalar and longitudinally polarized matrix elements are equal up to corrections of order M/E.

This argument requires that the leading term does not vanish, i.e., the contraction with the four-momentum p does not yield zero. In fact, this would be the case if the gauge symmetry was exact such that the current associated with it was conserved. However, since the symmetry is broken, the scalar matrix elements are nonzero. We can conclude that S-matrix elements involving Goldstone bosons and S-matrix elements involving longitudinally polarized vector bosons become equal in the high-energy limit, the corrections being suppressed by the factor M/E.

3.1.2 Quasielastic Scattering at Leading Order

The Goldstone-boson equivalence theorem is easily verified by explicit calculations using the Feynman rules derived from the chiral Lagrangian (2.80) and any parameterization of the Goldstone multiplet Σ. In applying it to scattering processes one should take some care that there is only a single scale in the problem. For instance, the theorem holds for $2 \to 2$ scattering at a fixed angle (such that s, t, u all are going to infinity), but not for $s \to \infty$ holding t fixed. In the forward region vector boson scattering amplitudes are dominated by photon and transversal W, Z exchange. The equivalence theorem has no prediction for these parts of the scattering amplitude.

Leaving this complication aside, one can measure scattering amplitudes of longitudinally polarized vector bosons, apply the equivalence theorem and derive the structure of Goldstone boson interactions at high energies. Vice versa, one can adopt a model for Goldstone boson scattering amplitudes at high energies and use it to predict the limiting behavior for S-matrix elements of longitudinally polarized vector bosons. The interactions of transversally polarized vector bosons will not be affected by the details of the electroweak symmetry breaking sector to leading order. So, projecting out the longitudinal states of W and Z bosons (which is done using angular distributions of their decay products) will always be advantageous for isolating signatures of electroweak symmetry breaking.

The lowest-order effective Lagrangian (2.80) provides a prediction for the quasielastic $2 \to 2$ scattering amplitudes of W and Z bosons. Projecting onto longitudinal states and taking the high-energy limit, one arrives at the *Low-Energy Theorem* (LET) [43]:

$$A(W_L^- W_L^- \to W_L^- W_L^-) = -\frac{s}{v^2}, \qquad (3.7)$$

$$A(W_L^+ W_L^- \to W_L^+ W_L^-) = -\frac{u}{v^2}, \qquad (3.8)$$

$$A(W_L^+ W_L^- \to Z_L Z_L) = \frac{s}{v^2}, \qquad (3.9)$$

$$A(Z_L Z_L \to Z_L Z_L) = 0. \qquad (3.10)$$

The cross sections for on-shell scattering are calculated by squaring the amplitudes, inserting phase space factors and dividing by a symmetry factor of two for like-sign W and for ZZ final states. This symmetry factor will not be included in the amplitude in any of the relations given here.

Using the equivalence theorem, the LET amplitudes (3.7–3.10) are easily obtained by expanding the mass-generating operator $-\frac{v^2}{4} \operatorname{tr} [V_\mu V^\mu]$ in terms of Goldstone fields, cf. (2.72). The quartic coupling has two derivatives in it which translate into factors of s, t, and u. The four-Z amplitude is proportional to $s+t+u$ which vanishes in the given approximation. A direct computation of the amplitudes in terms of vector bosons, taking the high-energy limit in the end, is more involved. There are contributions from the four-boson vertex as well as s-, t-, and u-channel diagrams with three-boson vertices. The superficial energy dependence for longitudinal scattering is s^2 in the individual Feynman diagrams, but the leading term cancels in the sum, leaving the same result as the leading high-energy behavior.

It is somewhat confusing that we have to take the high-energy limit to arrive at a low-energy theorem. Actually, the approximation is valid in an energy range where the W and Z masses are negligible but no additional degrees of freedom interacting with Goldstone bosons (e.g., the Higgs) are yet present, nor do the higher-order operators $\mathcal{L}_1$–$\mathcal{L}_{11}$ play a role. If there is a light Higgs boson, this range is nonexistent. If not, the range of validity can at most reach from the W mass scale up to the TeV range.

Nevertheless, the LET has nontrivial physical content. It predicts the numerical coefficient of the lowest-order term of an expansion in terms of $E/4\pi v$, corrections of order M_W/E neglected. Since $M_W = gv/2$, this is in fact the limit $g \to 0$ with v fixed (m_H fixed if a Higgs boson is present). The lower boundary of the validity range, M_W, vanishes in this limit.

The LET is completely determined by low-energy parameters. No assumption about the nature of symmetry breaking has to be made, since the low-energy effective theory is necessarily described by the chiral Lagrangian (2.80). Although there are anomalous couplings which affect the high-energy behavior of the scattering amplitudes (see below) they do not modify the LET since their contributions are of the order E^4.

The only assumption on which the LET is based is the possibility to write down a low-energy expansion, i.e., the reliability of the effective-theory approach. There is no question about that if quantum field theory is a good description of physics at these energies after all, and if we have included the complete accessible spectrum in the formulation of the effective Lagrangian. In practice, however, it is not easy to extract the E^2 piece of amplitudes measured by experiment and to separate it unambiguously from higher-order contributions (which could be large) since in reality it is not possible to vary the gauge couplings g and g'.

3.1.3 Custodial Symmetry Relations

Assuming that custodial $SU(2)_C$ symmetry (cf. Section 2.3.3) is a generic feature of electroweak symmetry breaking, one can derive relations among the quasielastic Goldstone amplitudes which hold beyond the LET, independent of the form of higher-order contributions. These relations are most useful in the case of the amplitudes becoming strong, since they are violated by small corrections proportional to g'^2 (the hypercharge coupling) and $(m_t^2 - m_b^2)/16\pi^2$ (the fermion mass splittings, which enter via loop diagrams).

The Goldstone bosons transform under $SU(2)_C$ transformations as a triplet. This is exactly analogous to low-energy pion interactions where the role of $SU(2)_C$ is played by isospin. Isospin conservation allows the expression of all quasielastic scattering amplitudes in terms of a single function $A(s, t, u)$:

$$A(W_L^- W_L^- \to W_L^- W_L^-) = A(t, s, u) + A(u, t, s), \tag{3.11}$$

$$A(W_L^+ W_L^- \to W_L^+ W_L^-) = A(s, t, u) + A(t, s, u), \tag{3.12}$$

$$A(W_L^+ W_L^- \to Z_L Z_L) = A(s, t, u), \tag{3.13}$$

$$A(W_L^- Z_L \to W_L^- Z_L) = A(t, s, u), \tag{3.14}$$

$$A(Z_L Z_L \to Z_L Z_L) = A(s, t, u) + A(t, s, u) + A(u, t, s). \tag{3.15}$$

The function $A(s, t, u)$ satisfies

$$A(s, u, t) = A(s, t, u). \tag{3.16}$$

As we have seen, its Taylor expansion begins with

$$A(s, t, u) = \frac{s}{v^2}. \tag{3.17}$$

In Section 2.3.5, we have argued that the low-energy effective theory may contain $SU(2)_C$-violating operators which are suppressed by a factor v^2/Λ^2, where Λ is a scale pertaining to new $SU(2)_C$-violating physics. These operators will break the $SU(2)_C$-relations among the Goldstone scattering amplitudes by corrections of the order v^2/Λ^2 and s/Λ^2. Furthermore, the relations are strongly violated in forward scattering where photon exchange is important. There, elastic WW scattering becomes singular while $WW \to ZZ$ stays finite.

3.1.4 Next-to-Leading Order Contributions

The one-loop corrections to the universal Goldstone boson scattering amplitude $A(s, t, u)$ can easily be computed. With a Higgs boson as a regulator, the result is for $s \ll M_H^2$ [44]

$$
\begin{aligned}
\mathrm{Re}\, A(s, t, u) = \frac{s}{v^2} &+ \frac{s^2}{16\pi^2 v^4}\left(\frac{1}{2}\ln\frac{M_H^2}{s} + \frac{9\pi}{2\sqrt{3}} - \frac{74}{9}\right.\\
&\left. + \frac{t(s+2t)}{6s^2}\ln\frac{M_H^2}{-t} + \frac{u(s+2u)}{6s^2}\ln\frac{M_H^2}{-u} - \frac{2(t^2+u^2)}{9s^2}\right)\\
&+ \alpha_4\frac{4(t^2+u^2)}{v^4} + \alpha_5\frac{8s^2}{v^4}.
\end{aligned}
\tag{3.18}
$$

Note that the non-logarithmic terms are numerically small. Neglecting them and replacing M_H by μ in a general renormalization scheme, one can rewrite this as [11, 45]

$$
\begin{aligned}
\mathrm{Re}\, A(s, t, u) = \frac{s}{v^2} &+ \frac{1}{16\pi^2 v^4}\left\{-\frac{(t-u)}{6}\left[t\ln\frac{-t}{\mu^2} - u\ln\frac{-u}{\mu^2}\right] - \frac{s^2}{2}\ln\frac{s}{\mu^2}\right\}\\
&+ \alpha_4\frac{4(t^2+u^2)}{v^4} + \alpha_5\frac{8s^2}{v^4}.
\end{aligned}
\tag{3.19}
$$

Without a specific regulator or renormalization condition, only the logarithmic terms can be unambiguously fixed. The dependence on the renormalization scale μ can be absorbed in a redefinition of the coefficients of the dimension-four operators $\mathcal{L}_4$ and $\mathcal{L}_5$:

$$\alpha_4(\mu) = \alpha_4(\mu_0) - \frac{1}{12}\frac{1}{16\pi^2}\ln\frac{\mu^2}{\mu_0^2}, \tag{3.20}$$

$$\alpha_5(\mu) = \alpha_5(\mu_0) - \frac{1}{24}\frac{1}{16\pi^2}\ln\frac{\mu^2}{\mu_0^2}. \tag{3.21}$$

These equations summarize the leading renormalization group effect on the anomalous couplings. The $SU(2)_C$-violating couplings $\alpha_{5,6,10}$ are scale independent to this order.

3.1.5 Unitarity Constraints

The optical theorem states that the total cross section for any process is equal to the imaginary part of the elastic forward scattering amplitude. If there is only elastic $2 \to 2$ scattering so that the imaginary part of the amplitude receives no contributions except from the total cross section for this scattering itself, this can be translated into a relation for the scattering amplitude $A(s, t, u)$. Expanding it in partial waves,

$$\begin{aligned} A(s,t,u) &= 32\pi \sum_\ell a_\ell(s)\,(2\ell+1)P_\ell(\cos\theta) \\ &= 32\pi \sum_\ell a_\ell(s)\,(2\ell+1)P_\ell(1+2t/s), \end{aligned} \tag{3.22}$$

which can be inverted to yield

$$a_\ell(s) = \frac{1}{32\pi}\int_{-s}^{0}\frac{dt}{s}A(s,t,u)\,P_\ell(1+2t/s), \tag{3.23}$$

each partial-wave amplitude a_ℓ has to satisfy

$$|a_\ell(s) - \mathrm{i}/2| = 1/2, \tag{3.24}$$

i.e., as a curve in the complex plane parameterized by s it has to stay on the *Argand circle*, a circle with radius $\frac{1}{2}$ around the point $\frac{\mathrm{i}}{2}$. In particular, the real part of the partial-wave amplitude can never exceed $1/2$.

   If there are inelastic channels accessible from a given initial state, to leading order these increase the imaginary part of the forward scattering amplitude but not the real part. In this case the amplitude will be inside the Argand circle. Assuming that there are no multi-particle final states, one can diagonalize the scattering amplitude matrix and find superpositions of external states which scatter only elastically. For these the relation (3.24) again has to hold exactly.

In the case of WW scattering amplitudes, some care must be taken in applying this reasoning. The forward region is included in the integral. From that region the integral receives a contribution due to transversal gauge boson exchange that is independent of g and g' and even diverges in the elastic-scattering case (Coulomb singularity). One has to cut off this region; this introduces an error in the determination of the partial-wave amplitudes.

Nevertheless, one can formally take the LET expressions (3.7–3.10), diagonalize the matrix of scattering amplitudes and project onto partial-wave amplitudes. Not surprisingly, the eigenamplitudes coincide with isospin eigenstates.

In terms of the amplitude function $A(s, t, u)$, the isospin eigenamplitudes are given by [46]

$$A(I = 0) = 3A(s, t, u) + A(t, s, u) + A(u, t, s), \qquad (3.25)$$

$$A(I = 1) = A(t, s, u) - A(u, t, s), \qquad (3.26)$$

$$A(I = 2) = A(t, s, u) + A(u, t, s). \qquad (3.27)$$

Inserting the LET expressions (3.7–3.10), one obtains

$$A(I = 0) = 2\frac{s}{v^2}, \qquad (3.28)$$

$$A(I = 1) = \frac{t - u}{v^2}, \qquad (3.29)$$

$$A(I = 2) = -\frac{s}{v^2}, \qquad (3.30)$$

which can expanded in terms of partial-wave amplitudes to yield the nonvanishing terms

$$a_{J=0}^{I=0} = \frac{s}{16\pi v^2}, \qquad (3.31)$$

$$a_{J=1}^{I=1} = \frac{s}{96\pi v^2}, \qquad (3.32)$$

$$a_{J=0}^{I=2} = -\frac{s}{32\pi v^2}. \qquad (3.33)$$

There is no higher spin involved if we remain with the LET amplitudes. The critical value $a = 1/2$ is reached at the energies

$$I = 0: \qquad E = \sqrt{8\pi}\, v = 1.2\,\text{TeV}, \qquad (3.34)$$

$$I = 1: \qquad E = \sqrt{48\pi}\, v = 3.5\,\text{TeV}, \qquad (3.35)$$

$$I = 2: \qquad E = \sqrt{16\pi}\, v = 1.7\,\text{TeV}. \qquad (3.36)$$

Obviously, at these energies the leading-order approximation breaks down.

Including higher-order corrections can shift the scale of unitarity saturation but will not change the overall picture. At an energy scale where the next-to-leading-order correction is sizable, higher orders will contribute as

much, and finite-order perturbation theory ceases to be predictive. In order
to have amplitude expressions that are at least in accord with unitarity beyond these scales, one can try to resum the perturbation series in a particular
way. The result depends on the chosen resummation prescription and does not
tell anything about the actual high-energy behavior. However, it can serve as
a consistent implementation of particular models with distinct features at
energies beyond the unitarity saturation threshold.

The idea of such unitarization models is to project each eigenamplitude
function $a_\ell(s)$ onto the Argand circle. Doing this, one assumes implicitly that
no new scattering channels are open, so that $2 \to 2$ quasielastic scattering
dominates at all scales.

1. The K-matrix unitarization model [47, 48] is not limited to the perturbative expansion. Assuming that $a(s)$ is a real-valued amplitude function
   one starts with, the unitarized amplitude is given by

$$a_K(s) = a(s) \frac{1 + ia(s)}{1 + a(s)^2}. \tag{3.37}$$

   Geometrically, the value $a_K(s)$ corresponds to the projection of the point
   $a(s)$ onto the Argand circle along the straight line connecting $z = a(s)$
   with $z = i$ (Fig. 3.1). This construction can also be applied if $a(s)$ is
   complex, but (3.37) becomes slightly more complicated in that case.

   By construction, the K-matrix prescription will never generate a resonance if there is none within the function $a(s)$. In this respect it can
   be regarded as a minimal unitarization model. In particular, if the LET
   expressions (3.31–3.33) are inserted, the function

$$a(s) = a_0 \frac{s}{v^2} \tag{3.38}$$

   translates into

$$a_K(s) = a_0 s \frac{v^2 + ia_0 s}{v^4 + a_0^2 s^2}. \tag{3.39}$$

   This unitarized amplitude will asymptotically approach the fixed point
   $a_K(s) = i$, a resonance at infinity. The poles are at $\pm v/\sqrt{a_0}$ on the imaginary axis, which formally corresponds to resonances with zero mass and
   large width.

2. To obtain the Padé unitarization model (also known as the inverse amplitude method) [49], one separates the amplitude into two pieces. Usually,
   one takes the leading term $a^{(0)}(s)$ and the real part of the next-to-leading
   order term $a^{(1)}(s)$ in the chiral expansion which are proportional to s and
   to s^2, respectively. Then, the unitarized amplitude reads

$$a_P(s) = \frac{a^{(0)}(s)^2}{a^{(0)}(s) - a^{(1)}(s) - ia^{(0)}(s)^2}. \tag{3.40}$$

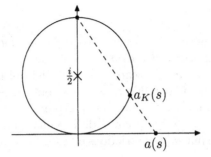

Fig. 3.1. K matrix construction for projecting a real scattering amplitude onto the Argand circle

If $a^{(1)}(s)$ vanishes, this coincides with the K-matrix model. However, if $a(s)$ has the form

$$a(s) = a_0 \left(\frac{s}{v^2} + \alpha \frac{s^2}{v^4} \right), \tag{3.41}$$

the Padé-unitarized amplitude is

$$a_P(s) = \frac{a_0 s^2/v^4}{s/v^2 - \alpha s^2/v^4 - i a_0 s^2/v^4}$$

$$= \frac{-a_0 s/\alpha}{s - v^2/\alpha + i a_0 s/\alpha}. \tag{3.42}$$

This is a resonance with mass and width

$$M = \frac{v}{\sqrt{\alpha}}, \quad \Gamma = \frac{a_0}{\alpha} M, \tag{3.43}$$

where the Breit-Wigner resonance shape is modified by a running-width prescription, which keeps the amplitude consistent with the LET. In other words, adopting the Padé unitarization method is equivalent to the assumption that each partial-wave amplitude in $2 \rightarrow 2$ scattering is dominated by a single resonance, an assumption that, incidentally, works reasonably well for pion-pion scattering in low-energy QCD. However, it need not be true for the scattering of electroweak Goldstone bosons.

3. There is an infinite number of possible unitarization models. An extreme case is the simple prescription

$$a_S(s) = e^{ia(s)} \sin a(s). \tag{3.44}$$

Inserting the LET amplitude, this results in an infinite series of resonances which become increasingly dense on the energy axis when s approaches infinity.

3.2 Resonances

A striking signature of new physics is a resonance in some scattering channel. If such resonances appear in Goldstone boson scattering, this would almost certainly give us a clue in the search for the origin of electroweak symmetry breaking. While some possible resonances could be elementary particles (such as a Higgs boson), others would be interpreted as bound states of more fundamental objects yet to be discovered. The quantum numbers present in Goldstone scattering put some restrictions on the resonances that can possibly be observable.

3.2.1 Resonance Multiplets

Particles in $SU(2)_L \times U(1)_Y$ multiplets are expected in weakly-coupled theories, where Σ couplings account for mass splittings within the multiplet after spontaneous symmetry breaking. First, we consider states with zero hypercharge. If they are to be seen as resonances in Goldstone scattering, they must be either singlets or triplets with respect to the left-handed $SU(2)_L$ symmetry. Assuming that $SU(2)_C$ is a good symmetry as well, this leaves the following possibilities for resonances:

$$\text{Scalar singlet } \sigma: \qquad \mathcal{L}_\sigma = g_\sigma \sigma \frac{v}{2} \operatorname{tr} \left[V_\mu V^\mu \right] \qquad (3.45)$$

$$\text{Vector triplet } \rho_\mu^a: \qquad \mathcal{L}_\rho = g_\rho \frac{v^2}{2} \operatorname{tr} \left[\rho_\mu^a \tau^a V^\mu \right] \qquad (3.46)$$

$$\text{Tensor singlet } \tau^{\mu\nu}: \qquad \mathcal{L}_\tau = g_\tau \frac{v}{2} \tau^{\mu\nu} \operatorname{tr} \left[V_\mu V_\nu \right] \qquad (3.47)$$

$$\vdots \qquad\qquad\qquad \vdots$$

In strongly coupled theories, a classification in terms of $SU(2)_C$ multiplets seems to be more appropriate, similar to low-energy QCD. Since for zero-hypercharge states the $SU(2)_L$ and $SU(2)_C$ quantum numbers coincide, (3.45–3.47) is simultaneously the list of $SU(2)_C$ multiplets that can be produced. The scalar and tensor are electrically neutral while the vector multiplet has a neutral and a charged component just as the $W^\pm, Z$ triplet does. As the expansion (2.72) shows, the triple vector coupling induced by (3.46) is antisymmetric, forbidding the coupling of a vector resonance to identical particles. It can decay into W^+W^- but not into ZZ.

For a resonance with mass M, in the $SU(2)_C$-symmetric case the amplitude functions $A(s, t, u)$ for a scalar and a vector resonance following from (3.45, 3.46) are

$$A_\sigma(s, t, u) = -g_\sigma^2 \frac{s^2}{v^4} \frac{1}{s - M^2}, \qquad (3.48)$$

$$A_\rho(s, t, u) = -g_\rho^2 \left(\frac{s - u}{t - M^2} + \frac{s - t}{u - M^2} + 3 \frac{s}{M^2} \right). \qquad (3.49)$$

which vanish proportional to s^2 in the low-energy limit. If $SU(2)_C$ violation is allowed, resonances transforming as scalar triplets, vector singlets, etc. are also possible, and in any case the amplitude relations between Z and W external states are lost.

By inserting Σ factors in the interaction operators, one can describe resonances with half-integer $SU(2)_L$ and unit hypercharge. One particular type consists of complex doublets which have four components each:

Scalar doublet $\tilde{\sigma}$: $\mathcal{L}_{\tilde{\sigma}} = g_{\tilde{\sigma}} \operatorname{tr} [\tilde{\sigma} \Sigma] \operatorname{tr} [V_\mu V^\mu]$ (3.50)

Vector doublet $\tilde{\rho}$: $\mathcal{L}_{\tilde{\rho}} = g_{\tilde{\rho}} \operatorname{tr} [\tilde{\rho}^\mu \Sigma V_\mu]$ (3.51)

$\vdots$ $\vdots$

Expanding this in terms of Goldstone fields and their derivatives, one observes that some members of the multiplets do not couple to Goldstone pairs. The multiplets decompose with respect to $SU(2)_C$ (a singlet and a triplet each), and only the states with the $SU(2)_C$ quantum numbers shown in (3.45–3.47) can actually be produced. This situation changes if one allows for $SU(2)_C$ violation in the couplings. Then, the remainder of the $SU(2)_L \times U(1)_Y$ multiplets can also show up in Goldstone scattering.

Apparently, the amount of information on new resonances accessible in Goldstone (longitudinal W/Z) scattering amplitudes is limited by custodial symmetry. Only if this symmetry is violated, one can hope to access complete multiplets of the electroweak symmetry. (Note that such couplings are usually suppressed in models with a linear Higgs representation.) Since one expects mass splittings within these multiplets caused by their couplings to the symmetry-breaking sector, it is important to check for the presence or absence of $SU(2)_L \times U(1)_Y$ partners and to compare their properties. The quantum numbers of these particles determine their couplings to transversal gauge bosons, so the obvious method to do this is by pair production in fermion annihilation. Since the photon coupling is of no use here once the charge is known, one has to pair-produce them from virtual Z and W bosons.

This discussion is not restricted to the nonlinear representation. In fact, in supersymmetric models where the light Higgs state makes a linear representation appropriate, an important experimental question is to disentangle the various Higgs states. There is a second Higgs doublet ($\tilde{\sigma}$ in our notation) which adds four physical scalar states, but there can also be Higgs singlets σ, welcome as a possible solution of the so-called μ problem. Concerning the second Higgs doublet, custodial symmetry would forbid couplings of the triplet part to Goldstone bosons. However, in supersymmetric models the alignment of the Higgs vacuum expectation values violates this symmetry and, furthermore, those states can be pair-produced due to their non-vanishing electroweak quantum numbers.

3.2.2 The Technirho

A vector resonance triplet ρ_μ^a typically appears in models of dynamical symmetry breaking [22, 50]. In such a scenario, it not meaningful to talk about its $SU(2)_L \times U(1)_Y$ quantum numbers, but one would rather classify it as a triplet with respect to the custodial $SU(2)_C$ symmetry only. For that reason, it is customary not to write down its Goldstone coupling in the form (3.46) but to use the formalism developed in Section 2.3.4.

For chiral transformations where the transformation matrix $C(x)$ becomes position-dependent, the composite vector field $\mathcal{V}_\mu$ (2.143) has a nonlinear transformation law (2.146), so a quadratic term $\text{tr}\,[\mathcal{V}_\mu \mathcal{V}^\mu]$ is not allowed by the symmetry. However, if we assign the following transformation law to $\rho_\mu \equiv \rho_\mu^a \tau^a$,

$$\rho_\mu \to C\rho_\mu C^\dagger + g^{-1}(\partial_\mu C)C^\dagger \tag{3.52}$$

with an arbitrary parameter g, the combination $(\mathcal{V}_\mu - g\rho_\mu)$ transforms linearly, and the quadratic form

$$
\begin{aligned}
&-a\frac{v^2}{4}\,\text{tr}\,[(\mathcal{V}_\mu - g\rho_\mu)(c\mathcal{V}^\mu - g\rho^\mu)] \\
&= -a\frac{v^2}{4}\,\text{tr}\,[\mathcal{V}_\mu \mathcal{V}^\mu] + ag\frac{v^2}{2}\,\text{tr}\,[\mathcal{V}_\mu \rho^\mu] - ag^2\frac{v^2}{4}\,\text{tr}\,[\rho_\mu \rho^\mu]
\end{aligned}
\tag{3.53}
$$

is chirally invariant. One should add a coupling to the electroweak gauge bosons,

$$\frac{v^2}{2}\,\text{tr}\,[\mathbf{W}_{\mu\nu}\rho^{\mu\nu}], \tag{3.54}$$

where $\rho^{\mu\nu}$ is the field strength tensor of the vector field ρ_μ. If $\mathcal{V}_\mu$ is expanded in terms of Goldstone fields, keeping only the lowest order, the first term on the right-hand side of (3.53) is a quartic Goldstone interaction, the second term is the usual vector coupling of scalar pairs, and the third term gives the ρ_μ boson a mass. From this, one can again derive the Goldstone scattering amplitude; it is identical to (3.49) with

$$M^2 = ag^2v^2 \quad \text{and} \quad g_\rho^2 = aM^2/4v^2. \tag{3.55}$$

For the parameterization to be useful in the resonant channels, the pole in (3.49) has to be regulated by inserting the width of the resonance. If there are no other decay channels than Goldstone bosons (i.e., longitudinal vector bosons), the width is given by

$$\Gamma = \frac{aM^3}{192\pi v^2} = \frac{g_\rho^2}{48\pi}M. \tag{3.56}$$

The width induces an imaginary part in the amplitude which should vanish like s^2 in the low-energy limit. The prescription for achieving this is described below for the case of a scalar resonance.

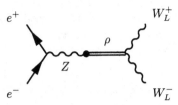

Fig. 3.2. Feynman graph for the mixing of a Z boson and a heavy vector resonance

In the formulation of Section 3.2.1, the bilinear coupling (3.46) of a vector triplet to the gauge bosons should, in unitary gauge, be removed by a field redefinition of ρ_μ and the electroweak gauge bosons. The mixing induces a common coupling of the ρ_μ resonance to all fermion fields. This is remarkable since otherwise these couplings are expected to be proportional to the fermion masses, negligible for the first generation. As a result, one expects a sensitivity to the ρ_μ resonance properties in the $e^+e^- \rightarrow W_L^+W_L^-$ scattering amplitude (Fig. 3.2). There is no contribution to the Z_LZ_L final state. Apart from affecting the measurement of triple gauge couplings (Section 4.1.3) in this way, in the low-energy effective Lagrangian vector states can provide direct contributions to the S parameter (Section 4.1.2) and thus modify the SM prediction for the Higgs mass derived from electroweak precision data.

3.2.3 The Higgs Resonance

The Higgs resonance is a $SU(2)_C$ singlet scalar which completes the $SU(2)_C$ triplet of Goldstone bosons to a complex doublet with respect to the full $SU(2)_L \times U(1)_Y$ symmetry. If and only if this resonance is present in Goldstone scattering, the scattering amplitude can be calculated reliably in perturbation theory and extrapolated to very high energies. The breakdown scale of the model is then set by a Landau pole in the Higgs self-coupling which appears when the perturbation series is resummed in the high-energy region. This is different from any other resonance case. Normally, the breakdown scale is of the order $4\pi v$, and higher-order calculations within a given model are of limited value.

In the following sections, we will therefore present explicitly the amplitude $A(s,t,u)$ for Goldstone scattering in the presence of a Higgs boson, calculated to leading order and resummed where it is necessary. Throughout this discussion we will stay in the limit $g, g' \rightarrow 0$ and ignore fermionic contributions. Numerically, this is appropriate only for fairly large Higgs masses ($\gtrsim 500\,\text{GeV}$), but it considerably simplifies the argument. If necessary, the gauge and fermionic corrections can be included. However, since the scattering of longitudinal W bosons has to be viewed as a subprocess of some fermionic scattering amplitude such as $e^+e^- \rightarrow$ six fermions, there are other contributions of order g, g' which have to be included as well and make these corrections non-universal.

In Chapter 2, the Higgs boson was introduced as a scalar coupled to the Goldstone field Σ either in the nonlinear representation

$$\Sigma = \left(1 + \frac{1}{v}H\right) \exp\left(-\frac{\mathrm{i}}{v} w_a \tau^a\right), \tag{3.57}$$

or in the linear representation

$$\Sigma = \left(1 + \frac{1}{v}H\right) - \frac{\mathrm{i}}{v} w_a \tau^a, \tag{3.58}$$

which describes the same physics but makes renormalizability manifest.

Actually, renormalizability is a property also of the nonlinear representation once the Higgs boson is included — i.e., when calculating loop diagrams, divergences will cancel in a way that is not obvious from power counting. One can find representations which interpolate between (3.57) and (3.58), for instance

$$\Sigma = \left[\left(\frac{1}{\eta} + \frac{1}{v}H\right) \exp\left(-\mathrm{i}\frac{\eta}{v} w_a \tau^a\right) + \left(1 - \frac{1}{\eta}\right)\right] \tag{3.59}$$

with a free parameter η. In the limit $\eta \to 0$, the linear representation (3.58) is reproduced while the choice $\eta = 1$ corresponds to the nonlinear representation (3.57). Representation independence will ensure that η drops out of perturbatively calculated scattering amplitudes when all contributing Feynman diagrams are summed together.

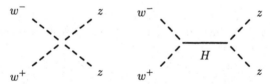

Fig. 3.3. Feynman graphs for Goldstone scattering in the presence of a Higgs resonance

When the Goldstone scattering amplitude $A(s,t,u)$ is calculated in a generic representation such as (3.59), one can identify two contributions, a contact term and a resonant Higgs exchange diagram (Fig. 3.3). The contributions of both diagrams individually depend on η, but in the sum the representation dependence drops out. To lowest order,

$$A^{(0)}(s,t,u) \equiv A^{(0)}(s) = -2\lambda \frac{s}{s - M^2} \tag{3.60}$$

with $\lambda = M^2/2v^2$. This expression has two important properties:

(a) At low energies, it approaches the LET value s/v^2. Looking at higher orders in the loop expansion, the sum of the diagrams in any fixed order

n of the perturbative expansion vanishes like $s^{n+1}/(4\pi v)^{2n} \log^n s$ for $s \to 0$. Thus, the LET low-energy behavior is not modified by higher-order corrections.

(b) At high energies, the amplitude $A^{(0)}(s)$ approaches a constant value. Higher-order corrections modify this by adding logarithmic terms of order $\lambda^n \ln^k s$, $k \leq n$.

The linear ($\eta = 0$) and the nonlinear ($\eta = 1$) representations are singled out. For $\eta = 0$ both the contact term and the resonant term approach a constant at low as well as at high energies. Thus, at high energies unitarity (i.e., the absence of contributions which increase more than logarithmically) is manifest while, to satisfy the LET, at low energies there is a large cancellation between the two contributions. Conversely, for $\eta = 1$ the cancellation occurs for high energies, while in the low-energy range both terms satisfy the LET individually. For that reason, one cannot drop the contact term even though it might be categorized as a background to Higgs production and decay.

This feature persists to all orders of perturbation theory. For avoiding large cancellations, the nonlinear representation is more appropriate for $s < M_H^2$ while one would like to choose the linear representation for $s > M_H^2$. The interpolating representation (3.59) comes handy here since one can define

$$\eta(s) = (1 + \sqrt{s}/M)^{-1}. \qquad (3.61)$$

This has the appropriate limits

$$\eta(0) = 1, \qquad \eta(\infty) = 0, \qquad (3.62)$$

such that large cancellations are avoided over the whole of phase space.

In the amplitude expression (3.60) the Higgs width has not yet been included. This must be done in a way that neither violates the LET nor the high-energy behavior [51]. In the high-energy limit one should resum furthermore the leading logarithms using renormalization group methods. More precisely, four energy ranges with different expansion parameters have to be distinguished:

1. The low-energy region ($s \ll M^2$): The amplitude is expanded in powers of $s/(4\pi v)^2$. The leading term and the imaginary part of the next-to-leading term are fixed by the low-energy theorem.

2. The perturbative region ($s \sim M^2$, but excluding the resonance region where $|s - M^2| \lesssim \lambda M^2/16\pi$): The amplitude is expanded in powers of $\lambda/16\pi = M^2/32\pi v^2$.

3. The resonance region ($|s - M^2| \lesssim \lambda M^2/16\pi$): The distance from the pole is of the order of the width $\Gamma \sim \lambda$. Any Feynman diagram contributing to $ww \to zz$ can be characterized by non-negative integers n and k to be of the order

$$16\pi \left(\frac{\lambda}{16\pi}\right)^n \left(\frac{\lambda M^2/16\pi}{s - M^2}\right)^k, \qquad n \geq 1, \quad k \geq 0, \qquad (3.63)$$

where k counts the number of resonant propagators, and all s-dependence that is not determined by the pole terms has been absorbed in the $k = 0$ piece. All terms with a fixed n are formally of the same order and need to be resummed.

4. The high-energy region $(s \gg M^2)$: Neglecting everything that is suppressed by powers of M^2/s, any term can be assigned non-negative integers n and k to be of the order

$$16\pi \left(\frac{\lambda}{16\pi} \right)^n \left(\frac{\lambda}{16\pi} \ln \frac{s}{M^2} \right)^k, \qquad n \geq 1, \quad k \geq 0. \tag{3.64}$$

All terms with a fixed n are formally of the same order and need to be resummed. This can be accomplished by renormalization-group methods, introducing a running coupling and field anomalous dimensions.

In each of these expansions the individual terms are representation independent. For the perturbative expansion in λ, this follows from the general theorems mentioned above. Regarding the expansion in powers and logarithms of s, the corresponding pieces can in principle be identified in a measurement of physical scattering processes.

The width enters the denominator of the amplitude if a Dyson resummation of the resonant part is performed. Separating the full amplitude into a resonant and a nonresonant part,

$$A(s, t, u) = A_{\text{res}}(s) + A_{\text{nr}}(s, t, u), \tag{3.65}$$

this resummation can be written in the general case as

$$A_{\text{res}}(s) = (\alpha + i\beta) \frac{M^2}{s - M^2} \sum_{k=1}^{\infty} \left(-i\frac{\Gamma}{M} \theta(s) \frac{M^2}{s - M^2} \right)^{k-1}$$

$$= (\alpha + i\beta) \frac{M^2}{s - M^2(1 - i\gamma)}, \tag{3.66}$$

where $\gamma \equiv (\Gamma/M)\theta(s)$, and $\alpha + i\beta$ is a complex constant.

In the lowest-order expression (3.60), if the resonant and nonresonant parts were identified with the corresponding Feynman diagrams, this splitting would become representation dependent. However, requiring M and Γ to represent the physical pole position in the complex plane, this ambiguity is removed. The complex constant $\alpha + i\beta$ denotes the residue of this pole, so this is representation independent as well. The remainder of the amplitude is shifted into A_{nr}. To lowest order, the result reads

$$A_{\text{res}}^{(0)}(s) = -2\lambda \frac{M^2}{s - M^2} \sum_{k=1}^{\infty} \left(-i\frac{\Gamma^{(0)}}{M} \theta(s) \frac{M^2}{s - M^2} \right)^{k-1}$$

$$= -2\lambda \frac{M^2}{s - M^2 + iM\Gamma^{(0)}}, \tag{3.67}$$

where

$$\Gamma^{(0)} = \frac{3\lambda}{16\pi} M = \frac{3g_\sigma^2}{32\pi} M \tag{3.68}$$

in the notation of (3.45). The nonresonant part is

$$A_{\mathrm{nr}}^{(0)}(s,t,u) = 2\lambda. \tag{3.69}$$

While this result restores unitarity in the resonance region, it violates the LET when both terms are added (3.65). The reason is the truncation of the perturbation series at finite order [52]. To remedy this, one should shift a constant from the nonresonant into the resonant piece (3.66):

$$\hat{A}_{\mathrm{nr}}(s,t,u) \equiv A_{\mathrm{nr}}(s,t,u) - \frac{\alpha + i\beta}{1 - i\gamma} \tag{3.70}$$

$$\hat{A}_{\mathrm{res}}(s) \equiv A_{\mathrm{res}}(s) + \frac{\alpha + i\beta}{1 - i\gamma}$$

$$= (\alpha + i\beta)\frac{1 + i\gamma}{1 - i\gamma} \frac{s}{s - M^2(1 + \gamma^2) + i\gamma s}. \tag{3.71}$$

If the truncated series is inserted in the sum

$$\hat{A}(s,t,u) = \hat{A}_{\mathrm{res}}(s) + \hat{A}_{\mathrm{nr}}(s,t,u), \tag{3.72}$$

the LET is restored without affecting the resonance in the given order of perturbation theory. Actually, to lowest order

$$\hat{A}(s,t,u) = -2\lambda\frac{s}{s - M^2 + i\frac{s}{M}\Gamma^{(0)}\theta(s)}. \tag{3.73}$$

The same expression with the running width in the denominator is obtained by unitarizing (3.60) using the K-matrix prescription. It also coincides with a naive Dyson resummation of the resonant Feynman diagram calculated in the interpolating representation (3.59, 3.61).

Finally, one has to consider a resummation of leading logarithms in the high-energy region. This is performed by adding the difference

$$\Delta A(s,t,u) = \left[A_{\mathrm{RG}}(s,t,u;\mu_0^2) - A_{\mathrm{HE}}(s,t,u;\mu_0^2)\right]\theta(s,t,u;\mu_0^2). \tag{3.74}$$

Here, the term A_{HE} is the amplitude calculated to fixed order in a high-energy effective theory where the Higgs mass is neglected. This amplitude contains an arbitrary renormalization scale μ_0; it is fixed by a matching condition

$$A_{\mathrm{he}}(s,t,u;M^2) = A_{\mathrm{HE}}(s,t,u;\mu_0^2), \tag{3.75}$$

where $A_{\mathrm{he}}(s,t,u;M^2)$ is the high-energy limit of the original amplitude:

$$A_{\mathrm{he}}(s,t,u;M^2) = \lim_{s \gg M^2} A(s,t,u). \tag{3.76}$$

The limit is to be taken such that constant terms and logarithms $\ln^k s$, $\ln^k u$, etc. are kept, but all terms suppressed by at least one power of $1/s$ are omitted. The matching point μ_0 is arbitrary; an explicit computation shows that

$$\mu_0^2 = \exp(2)\, M^2 \tag{3.77}$$

is a convenient choice.

The first term in (3.74) is calculated in the high-energy effective theory by renormalization group evolution from the scale μ_0, i.e., summing leading logarithms $\log s/\mu_0^2$. To exclude the forward and backward regions from this resummation (where it is not valid), we have introduced a cutoff function

$$\theta(s,t,u;\mu_0^2) = \theta(s - \mu_0^2)\,\theta(-t - \mu_0^2/2)\,\theta(-u - \mu_0^2/2); \tag{3.78}$$

for practical applications, some smooth cutoff might be more appropriate.

To lowest order, the improvement of (3.73) from renormalization-group evolution is given by the extra term

$$
\begin{aligned}
\Delta A(s,t,u) &= \left[A_{\mathrm{RG}}(s,t,u;\mu_0^2) - A_{\mathrm{HE}}(s,t,u;\mu_0^2) \right] \theta(s,t,u;\mu_0^2) \\
&= -2\lambda \frac{\beta_0\lambda \ln \frac{s}{\mu_0^2}}{1 - \beta_0\lambda \ln \frac{s}{\mu_0^2}} \theta(s,t,u;\mu_0^2),
\end{aligned} \tag{3.79}
$$

where the β function reads

$$\beta_0 = 12/16\pi^2. \tag{3.80}$$

This expression exhibits a Landau pole at

$$\sqrt{s} = \Lambda = \mu_0 \exp\left(\frac{1}{2\beta_0\lambda} \right). \tag{3.81}$$

The validity of perturbation theory is limited to energies below this scale [53].

While we have presented explicit results only for the lowest-order amplitude, the generic approach described here can be applied to any order in the perturbative expansion. Combining the results of NLO Feynman diagram calculations [54], the complete next-to-leading-order result has been derived in [51], cf. Fig. 3.4. Higher-order corrections can be computed as well [55], but turn out to be numerically important only for large Higgs masses where the scale Λ (3.81) is so close that perturbation theory breaks down anyway.

3.2.4 Multiple Higgs States

The previous discussion can easily be extended to the case of multiple Higgs resonances. The lowest-order amplitude for Goldstone scattering in the presence of several $SU(2)_C$ singlet scalars reads

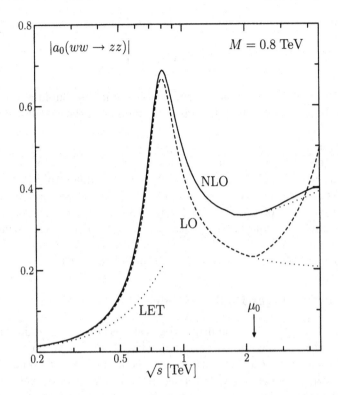

Fig. 3.4. Leading-order (LO) and next-to-leading-order (NLO) results for the Higgs lineshape. The plot shows the S-wave amplitude $a_0(s)$. The low-energy limit and the high-energy behavior without renormalization-group improvement are indicated by dotted lines [51]

$$A(s) = \frac{s}{v^2} - \sum_i \lambda_i \frac{s^2}{s - M_i^2} \qquad (3.82)$$

(compare (3.48)), where λ_i is the square of the Higgs coupling to a pair of longitudinal vector bosons. If this is to approach a constant at high energies, the couplings must satisfy the sum rule [56]

$$\sum_i v^2 \lambda_i = 1. \qquad (3.83)$$

This constant is given by

$$\lim_{s \to \infty} A(s) = -\sum_i \lambda_i M_i^2. \qquad (3.84)$$

For the theory to remain perturbative, one has to impose a unitarity constraint in the $I = J = 0$ channel (if $SU(2)_C$ is exact), which is equivalent to

$$|A(s)| < 8\pi \qquad (3.85)$$

away from the resonance regions. Thus, the Higgs masses have to fulfil the inequality

$$\frac{1}{8\pi} \sum_i \lambda_i M_i^2 < 1. \tag{3.86}$$

If a Higgs boson is heavy, in a renormalizable model its coupling to longitudinal vector bosons must be small. This is an example of the so-called *decoupling limit*.

Apparently, the presence of $SU(2)_C$ singlet scalars with appropriate couplings renders the model unitary, at least in the boson sector. In fact, this happens irrespective of the actual $SU(2)_L \times U(1)_Y$ quantum numbers of the scalars. As an example, consider a $SU(2)_L \times U(1)_Y$ singlet scalar field σ. If there is a coupling of the form (3.45) with the appropriate strength, there will be no unitarity violation in the leading-order Goldstone scattering amplitudes. Actually, this will cancel all divergences in bosonic one-loop diagrams. We will have a closer look at this scenario in Section 3.3.1.

3.2.5 The Two-Higgs-Doublet Model

The most popular model of multiple Higgs states contains two doublets [57]. In supersymmetric theories in particular [35], there can be no vertices that couple a single Higgs doublet both to up-type and down-type quarks, so one must introduce a second doublet. This is a straightforward explanation for the strong $SU(2)_C$ violation in the fermion sector. For a parameterization, we define two fields which transform as complex doublets under $SU(2)_L \times U(1)_Y$:

$$\Sigma_u = \frac{1}{v_u} \left[(v_u + \phi_u^0) - i\phi_u \right] \quad \text{and} \quad \Sigma_d = \frac{1}{v_d} \left[(v_d + \phi_d^0) - i\phi_d \right]. \tag{3.87}$$

where $\phi_{u,d} \equiv \phi_{u,d}^a \tau^a$.

The matrix notation will allow us to make the $SU(2)_C$ transformation properties explicit. Under $SU(2)_L \times U(1)_Y$ (or $SU(2)_L \times SU(2)_R$) transformations, both fields are assumed to have identical transformation properties:

$$\Sigma_{u,d} \to U \Sigma_{u,d} V^\dagger. \tag{3.88}$$

Both matrices are normalized to have unit expectation value,

$$\langle \Sigma_u^\dagger \Sigma_u \rangle = \langle \Sigma_d^\dagger \Sigma_d \rangle = 1, \tag{3.89}$$

where in perturbation theory we choose a gauge such that

$$\langle \Sigma_u \rangle = \langle \Sigma_d \rangle = 1. \tag{3.90}$$

The fermion mass terms are

$$\mathcal{L}_3 = -\bar{Q}_L M_{Qu} \Sigma_u \frac{1+\tau^3}{2} Q_R - \bar{Q}_L M_{Qd} \Sigma_d \frac{1-\tau^3}{2} Q_R$$

$$- \bar{L}_L M_{Qu} \Sigma_u \frac{1+\tau^3}{2} L_R - \bar{L}_L M_{Qd} \Sigma_d \frac{1-\tau^3}{2} L_R \qquad (3.91)$$

$$- \bar{L}_R^c M_{N_R} \frac{1+\tau^3}{2} L_R.$$

This particular form allows us to trade the large up-down mass splitting in favor of a large splitting between the two vacuum expectation values. If we have

$$\tan\beta \equiv v_u/v_d \gg 1, \qquad (3.92)$$

the entries of the dimensionless Yukawa coupling matrices,

$$\lambda_{Qu} = M_{Qu}/v_u \quad \text{and} \quad \lambda_{Qd} = M_{Qd}/v_d, \qquad (3.93)$$

can be similar in magnitude within each generation.

Vector bosons receive their masses from a combination of both vacuum expectation values:

$$\mathcal{L}_{2(W)} = \tfrac{1}{4} v_u^2 \operatorname{tr} \left[(D_\mu \Sigma_u)^\dagger D^\mu \Sigma_u \right] + \tfrac{1}{4} v_d^2 \operatorname{tr} \left[(D_\mu \Sigma_d)^\dagger D^\mu \Sigma_d \right]. \qquad (3.94)$$

Therefore,

$$v_u = v \sin\beta \quad \text{and} \quad v_d = v \cos\beta, \qquad (3.95)$$

such that

$$v_u^2 + v_d^2 = v^2. \qquad (3.96)$$

A direct contribution to the ρ parameter from dimension-four operators is absent as in the minimal model.

Analyzing (3.94), one concludes that the Goldstone bosons which get identified with the longitudinal vector boson states are given by

$$\mathbf{w} = \cos\beta \, \phi_u + \sin\beta \, \phi_d. \qquad (3.97)$$

The orthogonal linear combination,

$$\bar{\mathbf{w}} = -\sin\beta \, \phi_u + \cos\beta \, \phi_d, \qquad (3.98)$$

makes up a triplet of observable scalars, the $H^\pm$ and the A boson. In the limit of conserved $SU(2)_C$, these particles are degenerate in mass.

The two $SU(2)_C$ singlet states can mix, depending on the actual form of the Higgs potential. Denoting this mixing angle by α, one observes that the couplings of the mass eigenstates h and H to vector boson pairs, at tree level, are proportional to $\sin(\beta - \alpha)$ and $\cos(\beta - \alpha)$, respectively. If $\beta = -\alpha = \pi/2$, the tree-level coupling of the light Higgs boson vanishes. In this case, unitarity restricts the heavy Higgs mass to be less than about 1 TeV as in the minimal model. The light Higgs would be detectable, however, by its gluon and photon couplings which occur via one-loop heavy-fermion diagrams.

3.2.6 Resonance Decays

Any resonance that couples to longitudinal vector bosons is likely to have couplings to other particles as well. If kinematically allowed, the corresponding extra decay channels increase the resonance width. In the extreme case, the width is dominated by non-W/Z channels, such that quasielastic scattering of longitudinal vector bosons does not show resonant behavior at all. As an example, the Higgs resonance amplitude (3.73) assumes its maximum absolute value in the s channel at $s = M^2$,

$$|\hat{A}(s,t,u)|_{\max} = 2\lambda \frac{M}{\Gamma}, \tag{3.99}$$

which is equal to $16\pi/3$ if the Higgs couples only to Goldstone bosons, in which case the width is given by (3.68). This value becomes smaller if other decay channels are allowed:

$$|\hat{A}(s,t,u)|_{\max} = \frac{16\pi}{3} \, \text{BR}(H \to zz, ww). \tag{3.100}$$

If there are final-state particles with strong coupling to the Higgs scalar (or if there is a multitude of weakly-coupled particles), the resonance shape in quasielastic scattering is washed out and the amplitude merely approaches its high-energy asymptotic form as discussed before.

These final states could be searched for directly. They may carry a conserved quantum number which is not shared by any Standard Model particle. Then, they would not be observable but rather contribute to the dark matter content of the universe (if charge- and color-neutral). The neutralino of supersymmetric models is such a candidate. Another possibility is heavy-quark final states. In particular, if the Higgs mass is above the top pair threshold, there is a considerable (but subdominant) branching ratio for decays into $t\bar{t}$. Observing a heavy Higgs resonance in the scattering process $W^+W^- \to t\bar{t}$ with the expected strength would unveil an important piece of information, namely that masses in the fermion sector are generated by the same mechanism that is responsible for vector boson masses. Analogous considerations apply to the $b\bar{b}$ final state for a light Higgs boson.

Finally, there could be other decay channels which typically lead to four-fermion final states, similar to the longitudinal Z and W boson pairs of quasielastic scattering. An example are nonminimal Higgs models, where a heavy Higgs state could decay into light Higgses which in turn have fermionic decays. Here, the resonance strength is proportional to trilinear couplings, parameters in the Higgs potential.

To summarize, measuring all accessible two-particle decay channels of a Higgs resonance is a crucial part of the experimental program for establishing the origin of mass generation. If the minimal version of the Standard Model (SM) is correct, all Higgs couplings to particle pairs are proportional to the respective particle masses. By comparing branching ratios with this expectation, the nature of the Higgs boson as a quantum associated to scale fluctuations (in the SM, fluctuations of the vacuum expectation value v^2) can be established. To fully verify the SM expectation, one has to determine the absolute value of at least one Higgs partial width (or the total Higgs width), which requires cross section measurements in addition to branching ratio determinations. In addition, Higgs pair production has to be checked, as will be discussed in Section 3.3.1. Since there are good reasons to believe that the SM is not a complete theory of mass generation, precise measurements of these quantities are essential in searching for deviations from the SM predictions which give hints for the truly underlying theory.

These considerations are not restricted to Higgs bosons. No generic predictions exist for the couplings of non-scalar WW scattering resonances, so measuring them reveals new, completely independent information beyond the Standard Model. For instance, if there are composite resonances with confined constituents, fermionic decays could indicate mixing of constituents with standard fermions, providing important information about substructure.

3.3 Non-Resonant Inelastic Scattering

Measuring a resonance strength in production and decay gives access to some of the trilinear couplings of this state. However, for a complete understanding of electroweak symmetry breaking one also needs the values of other trilinear as well as quartic couplings of Higgs-sector states. To access those, one has to measure the non-resonant part of $W_L W_L \to XX$ and $Z_L Z_L \to XX$ scattering amplitudes, where XX denotes any (two-particle) final state with a significant coupling to the symmetry-breaking sector. Obviously, this is a much more delicate task, and background reduction plays an essential role. In this section we discuss some of the possible signals and their interpretation.

The parameterization of quasielastic non-resonant Goldstone scattering has been discussed before (Section 3.1). The quartic vector boson couplings introduced there are an important measurement irrespective of the existence of Higgs bosons. Clearly, without Higgs bosons this measurement appears both easier (because of the uncancelled cross sections) and even more interesting (because they give insight in non-perturbative dynamics).

3.3.1 Higgs Pairs

When Higgs bosons are present, the scattering matrix of the Goldstone triplet is extended. In the Minimal Standard Model, the Goldstone/Higgs scattering amplitudes read to leading order ($\lambda = M_H^2/2v^2$) [46]

$$A(w^+w^- \to w^+w^-) = -2\lambda \left(\frac{M_H^2}{s - M_H^2} + \frac{M_H^2}{t - M_H^2} + 2 \right), \qquad (3.101)$$

$$A(w^+w^- \to zz) = -2\lambda \left(\frac{M_H^2}{s - M_H^2} + 1 \right), \qquad (3.102)$$

$$A(zz \to zz) = -2\lambda \left(\frac{M_H^2}{s - M_H^2} + \frac{M_H^2}{t - M_H^2} + \frac{M_H^2}{u - M_H^2} + 3 \right), \quad (3.103)$$

$$A(w^+w^- \to HH) = -2\lambda \left(\frac{3M_H^2}{s - M_H^2} + \frac{M_H^2}{t} + \frac{M_H^2}{u} + 1 \right), \qquad (3.104)$$

$$A(zz \to HH) = -2\lambda \left(\frac{3M_H^2}{s - M_H^2} + \frac{M_H^2}{t} + \frac{M_H^2}{u} + 1 \right), \qquad (3.105)$$

$$A(HH \to HH) = -2\lambda \left(\frac{9M_H^2}{s - M_H^2} + \frac{9M_H^2}{t - M_H^2} + \frac{9M_H^2}{u - M_H^2} + 3 \right). \quad (3.106)$$

In the low-energy limit, the amplitudes not involving Higgs bosons reduce to the LET formulae (3.7–3.10). In the high-energy limit $|s|, |t|, |u| \gg M_H^2$ only the contact terms survive, and a larger symmetry emerges: The Higgs boson and the Goldstone bosons together make up a full $SU(2)_L \times SU(2)_R = O(4)$ multiplet.

One would like to verify or prove this behavior false by experiment. To fully determine the Higgs quantum numbers, one will have to extract the quartic couplings $HHww$ and $HHzz$ from the observed scattering amplitudes. Furthermore, the Higgs potential which determines the HHH and $HHHH$ couplings is involved here. In a generic model one expects deviations in all of these couplings which affect the above scattering amplitudes. This happens, for instance, if the Higgs sector is non-minimal, such that the lowest-lying state does not completely account for the vector boson and fermion masses.

As another example, the overall scale of electroweak interactions could be fixed by some dynamical mechanism, not directly related to electroweak symmetry breaking. In this case, scale fluctuations would be associated with a scalar σ transforming as a $SU(2)_L \times U(1)_Y$ singlet which couples to fermions and bosons proportional to their masses. Such a particle (a *dilaton* or *radion*) appears, for instance, in extra-dimension models [58]. Both the σ field and a Higgs boson, which might be present as well, transform as singlets under $SU(2)_C$ and are neutral, so they may mix.

All ratios of σ couplings to fermions and vector bosons, translating into ratios of branching ratios, may coincide with the SM expectations for a Higgs boson. For the Goldstone scattering amplitudes to be unitarized to first order, the existence of a $SU(2)_C$ singlet scalar with about the correct WW coupling strength is sufficient. This possibility essentially removes the upper limit on the Higgs boson mass resulting from unitarity.

It appears natural to use a linear parameterization for describing the scalar interactions even though the exact nature of the Higgs boson is not known beforehand. Universal deviations from the standard Higgs case can be parameterized in terms of three anomalous dimension-six operators [26, 27]:

$$\frac{v^2}{4} \left(\partial_\mu \operatorname{tr} \left[\Sigma^\dagger \Sigma \right] \right) \left(\partial^\mu \operatorname{tr} \left[\Sigma^\dagger \Sigma \right] \right),$$ (3.107)

$$\frac{v^4}{8} \operatorname{tr} \left[\Sigma^\dagger \Sigma \right]^3,$$ (3.108)

$$\frac{v^2}{4} \operatorname{tr} \left[(D_\mu \Sigma)^\dagger (D^\mu \Sigma) \right] \operatorname{tr} \left[\Sigma^\dagger \Sigma \right].$$ (3.109)

The first term is equivalent to a common shift in all Higgs couplings to matter such that their ratios are unaffected. Such a shift would first point to a model with a non-minimal Higgs sector. The second operator introduces a ϕ^6 term in the Higgs potential. After spontaneous symmetry breaking, it is equivalent to an anomalous triple Higgs coupling. Such a deviation is also generically expected in non-minimal Higgs models.

The operator (3.109) indicates a deviation in the Higgs quantum numbers. At least in weakly interacting models, the seagull interactions (the HHW^+W^- and $HHZZ$ vertices) are completely determined by gauge invariance. For a Higgs boson as a member of a complex $SU(2)_L \times U(1)_Y$ doublet, this interaction is given by

$$\frac{g^2}{4} W_\mu^+ W^{-\mu} H^2 + \frac{g^2}{8c_w^2} Z_\mu Z^\mu H^2,$$ (3.110)

regardless how many other Higgs states are present. On the other hand, for a pure singlet state these couplings may be absent. In the mixing case the normalization of the term (3.110) is free and translates into the coefficient of the dimension-six operator (3.109). Thus, the coefficient of this operator is the only unambiguous characteristic of the $SU(2)_L \times U(1)_Y$ quantum numbers of the Higgs boson.

If the observed scalar is assumed to be an electroweak singlet σ, the operators (3.107–3.109) correspond to new terms in the chiral Lagrangian

$$\sigma (\mathcal{L}_{2(W)} + \mathcal{L}_3),$$ (3.111)

$$\sigma^3,$$ (3.112)

$$\sigma \operatorname{tr} \left[V_\mu V^\mu \right],$$ (3.113)

where $\mathcal{L}_{2(W)}$ and $\mathcal{L}_3$ are the mass terms of the vector bosons and the fermions, respectively. In this case the scalar potential is not constrained by the requirement of electroweak symmetry breaking, thus the scalar self-coupling is not related to a ϕ^6 term.

Both effects, the presence of a singlet scalar and corrections to the Higgs potential, affect the Goldstone amplitudes $ww \to HH$ and $zz \to HH$. There are four Feynman diagrams (Fig. 3.5). Assuming that the Hzz coupling has been measured with sufficient accuracy in single Higgs production, the t- and u-channel contributions are known. If we denote deviations in the HHH and

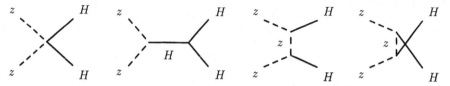

Fig. 3.5. Feynman diagrams for Higgs pair production in Goldstone scattering

$HHWW$ vertices by δ_{HHH} and δ_{HHWW}, respectively, the $W^+W^- \to HH$ amplitude reads

$$A(s,t,u) = -\frac{M_H^2}{v^2}\left((1+\delta_{HHWW}) + 3\frac{M_H^2(1+\delta_{HHH})}{s-M_H^2} + \frac{M_H^2}{t} + \frac{M_H^2}{u}\right).$$

(3.114)

At fixed energy, there is no way to disentangle the two anomalous couplings. (One would need the $HH \to HH$ amplitude for this task, which additionally contains the H^4 coupling.) So, only a certain linear combination can be measured. The best possible separation is achieved by varying the invariant mass of the Higgs pair between $s = 4M_H^2$ (the threshold) and infinity. The amplitudes are given by

$$A(4M_H^2) = -\frac{M_H^2}{v^2}\left(2 + \delta_{HHWW} + \delta_{HHH} - \frac{1}{\sin^2\theta}\right),$$

(3.115)

$$A(\infty) = -\frac{M_H^2}{v^2}\left(1 + \delta_{HHWW}\right),$$

(3.116)

where θ is the Higgs scattering angle in the zz c.m. frame.[1] While the seagull coupling can be deduced from the value of the amplitude at infinity, an unambiguous extraction of the triple Higgs coupling is possible only by subtracting the two amplitudes.

3.3.2 Fermion Pairs

While non-resonant quartic interactions of Goldstone and Higgs bosons are present in any model containing Higgs bosons, there is no need for contact vertices involving two Goldstone bosons and two fermions. In a linear representation, these correspond to dimension-five operators which are suppressed by one inverse power of a cutoff scale Λ.

Without Higgs bosons, a naive extrapolation of $ww \to \bar{f}f$ and $zz \to \bar{f}f$ to high energies yields an amplitude which eventually violates unitarity [59]. To be specific, the $I = J = 0$ partial-wave amplitude of Goldstone scattering into a f_+f_+ final state is given by

---

[1] Recall that in the forward region $\theta \approx 0$ the prerequisites of the equivalence theorem are not met, so the singularity at $\theta = 0$ should not be taken seriously.

$$a_0(s) = \frac{\sqrt{3N_c}}{16\pi v^2} m_f \sqrt{s}, \qquad (3.117)$$

where m_f is the fermion mass and $N_c = 3$ for quarks, $N_c = 1$ for leptons. This amplitude saturates the unitarity limit $\operatorname{Re} a_0 = \frac{1}{2}$ at the energy

$$\Lambda = \frac{8\pi v^2}{\sqrt{3N_c}\, m_f}. \qquad (3.118)$$

While for the top quark this value is around 3 TeV, not too far from the unitarity limit for $ww \to zz$, the scale is much higher for the other known fermions.

The rise of the amplitude can be compensated by dimension-five contact terms. Similar to the bosonic case, such operators also arise as counterterms in the one-loop corrections to the $ww, zz \to \bar{f}f$ amplitudes. For a closer analysis of fermion mass generation, it is worthwhile to study these processes and to extract the value of the anomalous couplings, in particular for the top quark.

There is a caveat here. The formula (3.117) is derived with the implicit assumption that the Higgs-less Standard Model is valid over the whole energy range. This cannot be true: The unitarity saturation in the bosonic sector necessitates a modification of the physics at or below 2 TeV, a scale below the unitarity limit for any fermion. If a Higgs boson appears before the unitarity limit is reached, its fermion couplings are likely to modify the previous conclusion.

On the other hand, if there is a Higgs boson which gives mass to the vector bosons, it does not necessary couple to fermions. If the Higgs-fermion couplings happen to vanish altogether, the unitarity limit (3.118) still applies, and there must be some new mechanism of fermion mass generation around or below this scale. Unfortunately, we do not know the structure of the theory beyond the unitarity limit in vector boson scattering, so the actual scale of fermion mass generation remains unknown. At best, we can hope that it is sufficiently low to be accessible in collider experiments.

3.3.3 New Particles

Apart from the known Standard Model particles, any new particle associated with electroweak symmetry breaking can likely be pair-produced in vector boson scattering. While pair production from electron and quark annihilation is a probe of the electroweak quantum numbers, pair production in longitudinal vector boson scattering is a sign of a close association with electroweak symmetry breaking. In particular, it provides a chance to observe particle multiplets which are electroweak singlets. In weakly interacting models of electroweak symmetry breaking, these signals are, unfortunately, very small and difficult to detect.

If the minimal supersymmetric extension of the Standard Model is correct [35], the particles directly associated with EWSB are the various Higgs states which give rise to both resonant and nonresonant contributions to $ww, zz \rightarrow XX$ amplitudes. Even if some of these particles escaped detection in electron or quark annihilation because of cancellations in their couplings to gauge bosons (due to mixing effects), they should be observable in vector boson scattering since they take part in the unitarization of the ww, zz scattering matrix. In addition, Higgs particles have fermionic partners (higgsinos) which mix with the superpartners of gauge bosons (gauginos) to give rise to charged and neutral Dirac and Majorana fermions (charginos and neutralinos). A measurement of the Higgs sector couplings of charginos and neutralinos would help to disentangle the gaugino and higgsino content of these states. Unfortunately, since all couplings in such models are typically small (of the order of gauge couplings), nonresonant contributions to cross sections are tiny and difficult to separate from transversal gauge boson interactions.

The situation may be more favorable if a dynamical mechanism of electroweak symmetry breaking is realized in Nature [36]. For instance, if due to extra spontaneously broken global symmetries several multiplets of pseudo-Goldstone bosons exist in the spectrum, their couplings to the known Goldstone bosons w, z may be sizable, at least at high energies. If such particles could be detected as final states in non-resonant vector boson scattering, valuable information beyond the quantum number assignment would be gained.

4

Collider Signatures

4.1 Precision Observables

The ultimate goal of phenomenological particle physics is the derivation of the fundamental degrees of freedom and the symmetries of their interactions from experimental observation. As far as the emerging theory allows for free parameters, they have to be determined as precisely as possible. They give no insight by themselves, but in searching for this theory, they represent all quantitative information that is available. Unless an ultimate parameter-free theory has been found, to make progress in that direction the best situation one can hope for is a non-renormalizable model where the next threshold of new physics is known. Today, non-renormalizability is present in gravitational interactions (corresponding to the Planck scale), in neutrino mixing (corresponding to an upper limit on the mass of right-handed neutrino states), and in electroweak interactions. In the latter case, the new physics threshold is within reach of the next generation of colliders, and there is actual hope that the current experimental results will soon be interpreted in terms of a theory of electroweak symmetry breaking.

However, until this stage has been reached, one has to draw conclusions from the quantitative results obtained within the context of the currently established (still Higgs-less) low-energy effective theory of electroweak interactions. One may assume a predictive underlying theory to be realized (such as the SM or a constrained version of the MSSM) and check this theory by comparing low-energy data with its predictions, but in a more phenomenological spirit we will adopt the generic effective-theory framework described in Chapter 2 and turn all experimental results into constraints of the coefficients in the effective action, calculated up to a certain order in a perturbative expansion.

4.1.1 Low-Energy Data

The interactions of light leptons provide some of the most stringent constraints of the electroweak theory. The $V - A$ structure in muon decay together with the neutral-current and charged-current neutrino interactions, the precision measurements of static electron and muon properties, and the absence of leptonic flavor-changing neutral currents firmly establish $SU(2)_L \times U(1)_Y$ and lepton family number (which is slightly violated by the neutrino mixing amplitudes) as symmetries of electroweak interactions. These results, confirmed by the leptonic interactions of real W and Z resonances, make the conclusion inevitable that there is a spontaneously broken $SU(2)_L \times U(1)_Y$ gauge symmetry. Unfortunately, the smallness of the lepton masses indicates a very weak direct coupling to the symmetry-breaking sector, so one has to study higher-order processes and observables to get a handle on it.

The quark interactions exhibit a more complicated pattern, but all observations so far are in accord with the same underlying structure of gauge interactions. Combining all semileptonic and nonleptonic data, everything seems to be consistent with the assumption of electroweak gauge interactions and a unitary CKM matrix which parameterizes the amount of quark family number violation.

To summarize, the detailed knowledge about low-energy interactions in conjunction with the discovery of the electroweak gauge bosons and the precise measurements of their properties uniquely fixes the structure of the electroweak effective Lagrangian as it has been developed in Chapter 2. The natural next step is a determination of the remaining coefficients in this Lagrangian, in particular those pertaining to operators which directly involve the symmetry-breaking sector. As we have argued, a direct probe is provided by the four-boson operators $\mathcal{L}_{4,5,6,7,10}$ in the list (2.88–2.98), but other structures are important as well. We will discuss them first.

4.1.2 S, T, and U

The observable scattering processes of lowest order in the electroweak coupling constants involve the exchange of at most one W or Z boson. The only effect of the symmetry-breaking sector on these processes is due to bilinear operators which contribute to the vector boson two-point functions. Vertex corrections and four-fermion operators can be neglected in this context (exept for heavy quark couplings) if one accepts the conjecture that the symmetry-breaking sector is coupled to light fermions proportional to their masses.

The bilinear couplings of vector bosons have already been discussed in Section 2.3.2,

$$\mathcal{L}_{WW} = \alpha_1 g g' \, \mathrm{tr} \left[\Sigma \mathbf{B}_{\mu\nu} \Sigma^\dagger \mathbf{W}^{\mu\nu} \right]$$
$$- \beta' \frac{v^2}{8} \, \mathrm{tr} \left[T V_\mu \right] \mathrm{tr} \left[T V^\mu \right] + \frac{1}{4} \alpha_8 g^2 (\mathrm{tr} \left[T \mathbf{W}_{\mu\nu} \right])^2. \tag{4.1}$$

These are operators of dimension two and four; the next terms in the expansion would be of dimension six, their coefficients suppressed by $1/\Lambda^2$, where Λ is the effective scale of electroweak symmetry breaking. If this scale is sufficiently high, the expansion can be truncated.

The coefficients α_1, β', and α_8 can be regarded as the leading coefficients in a Taylor expansion of that part of the gauge boson self-energies which is caused by new physics beyond the Standard Model. Of course, SM loop corrections give rise to contributions of similar structure. If no Higgs boson is present, this part is logarithmically divergent even after performing the renormalizations allowed by the gauge symmetry, and the first two of the above operators arise as extra counterterms. Once Higgs bosons are included, no such counterterms are needed and the logarithmic cutoff dependence is turned into a logarithmic Higgs mass dependence, but still any new physics will contribute a finite amount. (Typically, there are sizable contributions only to the first two operators.)

Peskin and Takeuchi have used this latter formulation to introduce the three parameters S, T, and U [15]. Looking at current correlators (which are equivalent to self-energies in the absence of vertex corrections), they define

$$S = 16\pi \frac{\partial}{\partial q^2} \left(\Pi_{33}(q^2) - \Pi_{3Q}(q^2) \right) \Big|_{q^2=0}, \tag{4.2}$$

$$T = \frac{4\pi}{s_w^2 c_w^2 M_Z^2} \left(\Pi_{11}(0) - \Pi_{33}(0) \right), \tag{4.3}$$

$$U = 16\pi \frac{\partial}{\partial q^2} \left(\Pi_{11}(q^2) - \Pi_{33}(q^2) \right) \Big|_{q^2=0}, \tag{4.4}$$

where the index $1, 3, Q$ indicates the first and third component of the weak current and the electromagnetic current, respectively. While in the original definition the correlators are evaluated at $q^2 = 0$, probing them at $q^2 \approx M_Z^2$ as it is done in actual experiments is equivalent to a shift in operator coefficients at the next higher order in the low-energy expansions, which we ignore here. The leading logarithmic corrections in the prefactors can be derived from the renormalization group and have to be included in a complete analysis [60].

Since without a Higgs state in the loop diagrams these coefficients are divergent, a reference Higgs mass is introduced, and only the shifts ΔS and ΔT with respect to the reference value are considered. Alternatively, one could use minimal subtraction or any other renormalization scheme to define the finite part of S and T. The Higgs mass cutoff is somehow physically motivated since a physical cutoff in the scalar channel is expected to unitarize WW scattering. Such a cutoff will also make S and T finite. If the actual Higgs mass is M_1, the shifts of the parameters with respect to a reference mass M_0 are as follows:

$$S(M_1) - S(M_0) = -\frac{1}{12\pi} \ln \frac{M_1^2}{M_0^2} + \cdots , \qquad (4.5)$$

$$T(M_1) - T(M_0) = \frac{3}{16\pi c_w^2} \ln \frac{M_1^2}{M_0^2} + \cdots , \qquad (4.6)$$

$$U(M_1) - U(M_0) = 0 + \cdots , \qquad (4.7)$$

where the dots indicate nonlogarithmic terms. In other words, increasing the reference Higgs mass reduces S and increases T. Conversely, if the actual Higgs mass turns out to be larger than expected, one needs a negative contribution to S and a positive contribution to T to compensate for this.

For $M \gg M_W$, only the logarithm $\ln(M_1/M_0)$ survives in the relations (4.5–4.7). For small Higgs masses the extra corrections must be taken into account. Clearly, if M_H is close to M_Z, higher-order terms in the Taylor expansion become important. The same would be true for new physics present at rather low scales, invalidating the truncation of the chiral expansion to some extent.

Keeping a possible renormalization scheme mismatch in mind (S, T, and U are defined with a Higgs cutoff at a scale M_H while the operator coefficients of the chiral Lagrangian are often defined in the $\overline{\text{MS}}$ scheme), the two parameter sets are related by [61]

$$\Delta S = -16\pi\alpha_1, \qquad (4.8)$$

$$\alpha\Delta T = -\beta', \qquad (4.9)$$

$$\Delta U = -16\pi\alpha_8, \qquad (4.10)$$

where α is the electromagnetic coupling $e^2/4\pi$.

All processes that do not involve more than single exchange of massive vector bosons depend on the Higgs mass only via these parameters. Leaving the Higgs mass free and ignoring the U parameter which plays only a minor role, for the Standard Model this leaves a single nontrivial constraint which can be checked by the abovementioned class of observables: Combining all precision data, some region in the ST plane will be singled out. The SM prediction yields a curve in the same plane along which the Higgs mass varies from its allowed lower bound up to the unitarity limit. This curve should cross the allowed region, if the Standard Model in its minimal form is valid.

Looking at the current status of precision data (Fig. 4.1), all results are actually consistent with each other when displayed in the ST plane. (An inconsistency at that point would indicate a breakdown of the chiral expansion, which is only possible if new physics exists at rather low scales.) The point $S = T = 0$ is inside the 90 % exclusion contour of the SM if the Higgs mass is assumed to be close to its lower limit. This is all the information that can currently be gained about the Higgs mass within the Standard Model context. One concludes that if there is a light Higgs boson, no sizable extra contributions from new physics to S and T are necessary to make theory and data consistent with each other.

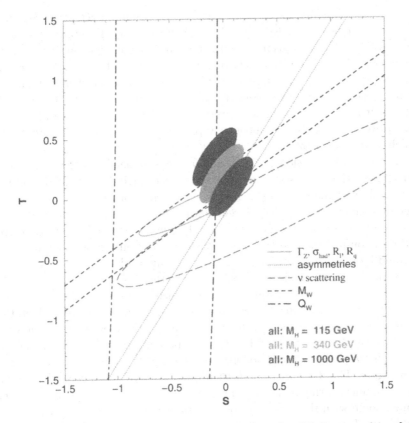

Fig. 4.1. 1σ constraints (39.35 % confidence level) in the ST plane resulting from the electroweak precision data. The lower, middle, and upper 90 % exclusion contours (filled) assume $M_H = 115\,\mathrm{GeV}$, $340\,\mathrm{GeV}$, and $1000\,\mathrm{GeV}$, respectively. $U = 0$ has been assumed in all fits (from [62])

Conversely, if no light Higgs exists, there must be such extra contributions to achieve agreement. Due to the nonrenormalizability of the model one naturally expects such contributions from physics at the cutoff scale. As a rough estimate, a correction to α_1 from physics at a scale Λ, for instance, is of the order of magnitude

$$\Delta\alpha_1 \sim \frac{v^2}{\Lambda^2}. \tag{4.11}$$

If $\Lambda \sim 4\pi v$, the natural upper limit for Λ in the absence of the Higgs resonance, this correction could compensate the shift from a light to a heavy Higgs boson (4.5). If there was a Higgs boson in the intermediate mass range, one would expect a higher cutoff scale and a smaller correction, while the shift due to the Higgs mass value would be diminished as well. Apparently, the precision data by themselves do *not* prefer any particular Higgs mass value.

However, the shift has to be in the right direction in the ST plane. Scanning a list of models, one may check the sign of ΔS and ΔT in any particular scenario [63]. To get agreement with data for a large Higgs mass value, one needs negative ΔS, positive ΔT, or a combination of the two. Nonzero ΔT corresponds to a certain amount of $SU(2)_C$ violation, which is not unlikely to be present if there is a connection to the fermion sector. Negative ΔS can result from an extra massive vector boson. On the other hand, QCD-like compositeness models (technicolor) typically predict a positive shift in ΔS which is disfavored by data.

The present precision of the S and T measurements is essentially a consequence of the large statistics collected at the LEP 1 experiments. Future collider data at higher energies are not expected to significantly improve this precision, unless a new, even larger data sample on the Z pole can be obtained. There is a project for doing this at a linear e^+e^- collider with electron and positron polarization (*Giga-Z*). With that option the precision on S and T could be improved by almost an order of magnitude [64].

Note that such precision, which requires the inclusion of leading NNLO corrections in the theoretical calculation, also necessitates the inclusion of the next order in the chiral expansion. The additional parameters which then have to be introduced remove the redundancy in the measurement. In a plot such as Fig. 4.1, the allowed bands corresponding to individual observables (M_W, $\sin^2 \theta_w$, Γ_Z, ...) need no longer intersect each other in a common exclusion region. The discrepancy can then be attributed to higher-order anomalous couplings. If the models in question allow for theoretical predictions that can compete with this level of experimental accuracy, the new measurements would be of great value for distinguishing among alternatives and pinning down their parameters.

4.1.3 Triple Gauge Couplings

The operators $\mathcal{L}_2$, $\mathcal{L}_3$, $\mathcal{L}_9$, and $\mathcal{L}_{11}$ in the list (2.88–2.98) do not contribute to vector boson two-point functions, but they modify the triple-gauge couplings. The bilinear operators discussed in the preceding section contribute to those as well. We will assume that that their values are known with sufficient precision to include them in the SM prediction and ignore their presence in this section. (Note that, in the SM part, there is no gauge-invariant separation between two-point, three-point, and higher vertex functions. This separation is meaningful only for the corrections due to new physics at higher scales which can be Taylor expanded and parameterized in terms of tree-level coefficients of a certain operator basis.)

Imposing CP invariance for simplicity, a standard parameterization of triple-gauge couplings is given by the operators [65]

$$\mathcal{L}_{WWV} = g_{WWV}\left[ig_1^V V_\mu (W^{-\nu}W^+_{\mu\nu} - W^-_{\mu\nu}W^{+\nu}) + i\kappa_V W^-_\mu W^+_\nu V^{\mu\nu}\right.$$

$$\left. + i\frac{\lambda_V}{M^2_W}W^{-\nu}_\mu W^{+\rho}_\nu V_\rho{}^\mu + g_5^V \epsilon^{\mu\nu\rho\sigma}(W^-_\mu \partial_\rho W^+_\nu - \partial_\rho W^-_\mu W^+_\nu)V_\sigma\right],$$

$$(4.12)$$

where V stands for the photon (γ or A_μ) and for the Z boson, and the pref-actors are $g_{WW\gamma} = e$ and $g_{WWZ} = gc_w$. Note that $g_5^{\gamma,Z}$ violates parity and charge-conjugation invariance but conserves CP. Parity is not a symmetry of electroweak interactions, so there is a priori no reason to expect this coefficient to be particularly suppressed. The SM values are

$$g_1^\gamma = g_1^Z = \kappa_\gamma = \kappa_Z = 1, \quad g_5^\gamma = g_5^Z = \lambda_\gamma = \lambda_Z = 0, \qquad (4.13)$$

such that only the shifts $\Delta g_1^{\gamma,Z}$ and $\Delta \kappa_{\gamma,Z}$ with respect to this value should be called anomalous couplings.

In principle, all coefficients are functions of the momentum transfer q^2. However, their momentum dependence can be unambiguously determined only within a specific scheme for separating propagator and vertex corrections. If we use the coefficients S, T, U for parameterizing "propagator-type" corrections, this scheme is determined only to leading order in an expansion in terms of external momenta, and for consistency we have to truncate the coefficient functions $g_1^{\gamma,Z}$, $\kappa_{\gamma,Z}$, $g_5^{\gamma,Z}$ and $\lambda_{\gamma,Z}$ at leading order as well.[1] Then, electromagnetic gauge invariance implies that

$$\Delta g_1^\gamma = g_5^\gamma = 0. \qquad (4.14)$$

The anomalous couplings $\Delta \kappa_{\gamma,Z}$, Δg_1^Z and g_5^Z are related to the operator coefficients in the chiral Lagrangian (2.80, 2.88–2.98) by [61]

$$\Delta \kappa_\gamma = g^2 \alpha_2 + g^2 \alpha_3 + g^2 \alpha_9, \qquad (4.15)$$

$$\Delta \kappa_Z = -g'^2 \alpha_2 + g^2 \alpha_3 + g^2 \alpha_9, \qquad (4.16)$$

$$\Delta g_1^Z = \frac{1}{c_w^2}g^2 \alpha_3, \qquad (4.17)$$

$$g_5^Z = \frac{1}{c_w^2}g^2 \alpha_{11}. \qquad (4.18)$$

Recalling that $\mathcal{L}_9$ and $\mathcal{L}_{11}$ violate the custodial symmetry, if we assume $SU(2)_C$ to be exact we get the further relations

$$\Delta \kappa_\gamma = -\frac{c_w^2}{s_w^2}(\Delta \kappa_Z - \Delta g_1^Z) \quad \text{and} \quad g_5^Z = 0, \qquad (4.19)$$

---

[1] This does not apply to SM radiative corrections which contain infrared logarithms from soft photon emission: These contributions cannot be expressed in the language of anomalous couplings and have to be taken into account separately.

so there are only two independent anomalous couplings which can be taken as $\Delta\kappa_Z$ and Δg_1^Z.

At this order in the chiral Lagrangian there is no operator generating an anomalous coupling λ. This parameter corresponds to dimension-six operators of the structure

$$\mathrm{tr}\left[\mathbf{W}_\mu^\nu\mathbf{W}_\nu^\rho\mathbf{W}_\rho^\mu\right], \tag{4.20}$$

which involve transverse gauge bosons only.[2] Thus, the prefactor $1/M_W$ in the definition of the λ coupling (4.12) should rather be read as

$$\frac{\lambda}{M_W^2} = \frac{1}{\Lambda^2}\left(\frac{\Lambda^2}{M_W^2}\lambda\right) = \frac{\lambda'}{\Lambda^2} \tag{4.21}$$

with λ' being, at most, of order one: The measurable value of λ has an inherent suppression factor M_W^2/Λ^2 where Λ is the typical scale of physics contributing to this term. On the other hand, the dimension-four operators $\mathcal{L}_{2,3,9,11}$ involve longitudinal gauge bosons, indicating a possible connection to the Higgs sector, and they effectively become dimension-six operators in the linear representation applicable above a possible Higgs mass threshold. Consequently, one expects their coefficients to carry an implicit suppression factor of just M_H^2/Λ^2 if the Higgs boson is not light, and one should neglect λ compared to these couplings in the parameterization of the triple gauge vertices. In the light-Higgs case, all couplings are of equal importance.

In a nonlinear parameterization (i.e., below the Higgs threshold), the coefficients of the $SU(2)_C$-symmetric dimension-four operators $\mathcal{L}_2$ and $\mathcal{L}_3$ are divergent. Similar to S and T, they may be defined using a reference Higgs mass on which they depend logarithmically, and new-physics contributions may be expressed in terms of the shift in those parameters with respect to the reference values. Within the SM, this provides a sensitivity to the Higgs mass which is independent of the information contained in S and T. However, at the time when these measurements can be performed to the required level of accuracy, a Higgs boson would have shown up directly, and precision measurements of the triple-gauge couplings would rather be used to look for physics beyond the Standard Model.

With today's data the level of accuracy in the determination of the triple-gauge couplings is much worse than for S, T, and U. The first meaningful measurements have been performed in W pair production at LEP 2 and at the Tevatron. Although one-loop radiative corrections need to be included when comparing theory and data, one is far from being sensitive to the Higgs mass value in this new channel. A much better accuracy is expected for LHC, and the limits can be further improved up to a similar level as S, T, U today if a high-energy, high-luminosity e^+e^- collider is available. This improvement is partly due to the increased luminosity, but even more a consequence of the higher energy available [66, 64].

---

[2] In the custodial-symmetric case (4.20) is in fact the only operator of this structure, and one can derive the relation $\lambda_\gamma = \lambda_Z$.

At e^+e^- colliders, the main channel for measuring these couplings is

$$e^+e^- \to W_L^+ W_L^- (\to 4f), \qquad (4.22)$$

which is equivalent, at high energies, to Goldstone pair production

$$e^+e^- \to w^+ w^-. \qquad (4.23)$$

The ZWW and γWW vertices are probed here at the full collider energy. The cross section for this process falls off like $1/s$, a consequence of gauge invariance which manifests itself as a cancellation between s-channel and t-channel diagrams. This behavior is not shared by the anomalous contributions: Each of the operators $\mathcal{L}_{2,3,9,11}$ involves two factors of the longitudinal vector field $V_\mu = \Sigma(D_\mu\Sigma)^\dagger$. In the calculation, the derivative couplings of the Goldstone bosons (the longitudinal polarization vectors of the physical W bosons) provide a factor of s which cancels the $1/s$ suppression.

Additional information can be obtained from single W, single Z, and single photon production,

$$e^+e^- \to e^\pm \nu_e W^\mp, \qquad (4.24)$$

$$e^+e^- \to \nu_e \bar\nu_e Z, \qquad (4.25)$$

$$e^+e^- \to \nu_e \bar\nu_e \gamma, \qquad (4.26)$$

where, independent of the energy, in most events the couplings are probed at a rather low scale of the order of the W mass. To get access to higher scales where the effect of anomalous couplings is more pronounced, one has to concentrate on the high-$p_\perp$ tail in the vector boson distributions.

4.2 *W* and *Z* Scattering Amplitudes

As we have discussed in detail in Chapter 3, the most direct probe of the symmetry-breaking sector is given by measurements of the scattering amplitudes $W_L W_L, Z_L Z_L \to XX$ of longitudinally polarized vector bosons into various final states. In the high-energy limit, these become equal to the scattering amplitudes of Goldstone bosons.

4.2.1 Phenomenology

There are three different ways to access such amplitudes in lepton and hadron colliders.

1. First of all, the imaginary part of the one-loop amplitude for the production of a final state XX (where $X = W, Z, t, \ldots$) is affected by a *rescattering correction*:

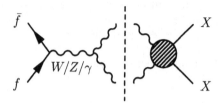

In this process, rescattering of the vector bosons takes place at the full c.m. energy of the annihilating fermions. This makes it the highest-energy probe of Goldstone scattering available. However, the imaginary part of an amplitude is difficult to extract. A realistic analysis will require a global fit of all parameters that describe particle pair production [67]. This includes, for the W^+W^- final state, the anomalous triple-gauge couplings described above. The quartic couplings that determine the imaginary part at next-to-leading order are just a small correction. Similar arguments apply to other final states like $t\bar{t}$.

Within a specific model, such an extraction nevertheless makes sense. Since the scattering is initiated by an (off-shell) spin-1 particle, the only available channel here is $J = 1$. While it is probably hopeless to look for the effects of genuine quartic couplings, a massive vector resonance (a technirho, see Section 3.2.2) will have a significant effect. In that case, the process is effectively a probe of the form factor in a vector-dominance situation, where both the real and the imaginary part carry information. For instance, with realistic assumptions about energy, luminosity, and detector performance, in $e^+e^- \rightarrow W^+W^-$ one is able to probe ρ_T masses significantly beyond the reach for direct production (cf. Fig. 4.2).

2. The second class of processes which are sensitive to the symmetry-breaking sector is W/Z associated pair production of a pair of particles XX:

$$f\bar{f} \rightarrow ZXX \tag{4.27}$$
$$f\bar{f}' \rightarrow WXX \tag{4.28}$$

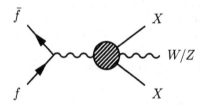

Since the intermediate W/Z line is far off-shell, transversal and longitudinal degrees of freedom mix, and the process gets a contribution from longitudinal vector boson scattering, where one vector boson is crossed into the final state.

Gauge invariance makes the cross section for all processes of this type fall off with $1/s$. Therefore, the best measurements are not necessarily done at the highest energy but somewhat above threshold. The variable

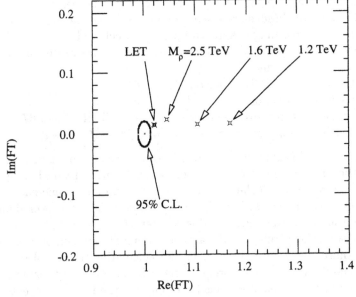

Fig. 4.2. Real and imaginary part of a vector-resonance dominated form factor for various values of the resonance mass. The ellipse is the 95 % exlusion contour for an assumed luminosity of 500 fb^{-1} at an e^+e^- collider with $\sqrt{s} = 500$ GeV (from [68])

to project out resonances or to observe the effect of anomalous couplings is the XX invariant mass. At higher energies, the vector boson fusion processes described below become more relevant.

This class of processes includes, for instance, the following:

- Higgs-strahlung: $f\bar{f} \rightarrow Zb\bar{b} \rightarrow f'\bar{f}'b\bar{b}$.
- Double Higgs-strahlung: $f\bar{f} \rightarrow ZHH \rightarrow 6f$
- Triple-vector boson production: $f\bar{f} \rightarrow ZW^+W^-, ZZZ \rightarrow 6f$. This is relevant both for Higgs physics (the decays $H \rightarrow WW^{(*)}$ and $H \rightarrow ZZ^{(*)}$) and for observing strongly interacting W bosons.

3. If the collider energy is sufficient, vector boson fusion becomes directly observable as a subprocess:

$$f\bar{f} \rightarrow f\bar{f}XX \tag{4.29}$$

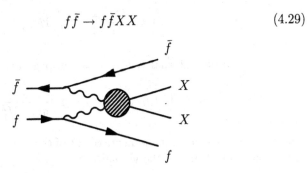

This class of processes has a cross section that rises logarithmically with energy. In the asymptotic limit where masses can be neglected, the intermediate vector bosons are essentially on-shell, and the amplitudes described in the preceding chapter are probed directly, albeit at a reduced c.m. energy. The processes that can be observed are analogous to the previous list, for instance:

- Higgs production in WW fusion: $f\bar{f}' \to f\bar{f}'b\bar{b}$
- Higgs pair production: $f\bar{f}' \to f\bar{f}'HH \to 6f$
- quasielastic vector boson scattering: $f\bar{f}' \to f\bar{f}'WW, ZZ \to 6f$
- Vector boson fusion into top quark pairs: $f\bar{f}' \to f\bar{f}'t\bar{t} \to 8f$

While the overall behavior of the amplitudes is easier to discuss in the Goldstone-boson picture (which corresponds to the limit of vanishing gauge couplings), for realistic calculations in a collider environment one has to use the full electroweak model with all degrees of freedom. This has been done to obtain the results described later in this chapter.

Furthermore, the complex final states consisting of four, six, or even eight fermions require an elaborated machinery for calculating the complete signal and background, even to lowest order in perturbation theory, if a reasonable accuracy is desired. Automated tools for evaluating the matrix elements and integrating over phase space are essential. Many of the results presented in the following sections have been obtained by means of such matrix element calculation, integration, and event generation packages [69, 70, 71, 72].

4.2.2 The Effective W Approximation

Looking at the structure of the Feynman graphs contributing to vector boson fusion processes, it is straightforward to derive the *effective W approximation* (EWA) which is valid in the high-energy limit [73], an approximation similar to the Weizsäcker-Williams approximation for photons (effective photon approximation, EPA [74]). This allows for a quick estimate of particle production cross sections initiated by W/Z pairs, although for a detailed quantitative analysis it is less useful, for reasons discussed below.

The matrix element in the cross section formula for the complete process $f_1 f_2 \to f_1' f_2' X$ with massless fermions,

$$\sigma(s) = \frac{(2\pi)^4}{2s} \frac{1}{4} \sum_{\text{pol}} \int d\Phi |T_{fi}|^2, \tag{4.30}$$

is factorized at one internal vector boson propagator:

$$T_{fi} = \sum_\lambda \left(\bar{u}(p)\not{\epsilon}_\lambda^*(v - a\gamma_5)u(p') \frac{1}{q^2 - M_V^2} \right) (T_{fi}')_\lambda, \tag{4.31}$$

where p and p' are the four-momenta of one of the incoming fermions, $q = p - p'$ is the four-momentum of the virtual vector boson ($q^2 < 0$), M_V is the vector boson mass, and v, a are the vector and axial-vector couplings of the incoming fermions. $(T'_{fi})_\lambda$ is the subprocess matrix element for the scattering of a λ-polarized virtual vector boson, where in the sum λ runs over all polarization directions.

The structure of the propagator is such that, at high energies, only small values of $|q^2|$ (of the order of M_V^2) contribute. If the reduced matrix element T'_{fi} is slowly varying on the scale of M_V^2, one can replace it by its on-shell value for $q^2 = M_V^2$. The polarization vectors are replaced by the on-shell polarization vectors for a particle with four-momentum q.

Furthermore, in this limit where q^2 is small, the momentum of the final-state fermion p' is approximately proportional to p with a small transverse momentum,

$$p' = (1 - x)p - q_\perp, \tag{4.32}$$

where $0 < x < 1$. The approximation now consists of factorizing the integration over $q_\perp$, which is possible in the small-angle limit where the dependence of the subprocess kinematics on $q_\perp$ can be neglected. Since typical scattering angles are, in the c.m. frame, proportional to

$$\sin\theta \approx \frac{|q_\perp|}{E(p')} \sim \frac{M_W}{(1-x)\sqrt{s}}, \tag{4.33}$$

this assumption is valid if $s \gg M_W^2$ and if x is not too close to one. Note that for $x \to 0$ the approximation also breaks down because for low subprocess energy the matrix element can no longer be treated as slowly varying.

The result takes the form

$$\sigma = \sum_\lambda \int \mathrm{d}x\, F_\lambda(x)\, \sigma_\lambda, \tag{4.34}$$

where the cross section for the subprocess initiated by the intermediate vector boson is given by

$$\sigma_\lambda = \frac{(2\pi)^4}{2xs}\frac{1}{2}\sum_{\mathrm{pol}}\int \mathrm{d}\Phi'|(T'_{fi})_\lambda|^2. \tag{4.35}$$

The probability $F_\lambda(x)$ for a polarized vector boson with momentum fraction x being emitted from the initial fermion can be evaluated for transversal and longitudinal polarization directions, respectively, to yield

$$F_\pm(x) = \frac{v^2 + a^2}{8\pi^2}\frac{1 + \bar{x}^2}{x}\ln\left(\frac{p_\perp^2 + \bar{x}M_V^2}{\bar{x}M_V^2}\right), \tag{4.36}$$

$$F_0(x) = \frac{v^2 + a^2}{4\pi^2}\frac{\bar{x}}{x}\frac{p_\perp^2}{p_\perp^2 + \bar{x}M_V^2}, \tag{4.37}$$

with $\bar{x} \equiv 1 - x$ and $p_\perp$ being a cutoff on the transverse momentum $q_\perp$.

Inserting the couplings of W and Z bosons, for a transverse cutoff $p_\perp \gg M_{W,Z}$, this reduces to

$$F_{W,\pm}(x) = \frac{g^2}{8\pi} \frac{1+\bar{x}^2}{x} \ln \frac{p_\perp^2}{M_W^2}, \tag{4.38}$$

$$F_{Z,\pm}(x) = \frac{g^2}{8\pi c_w^2} \left[(t_3^f - 2q_f s_w^2)^2 + (t_3^f)^2 \right] \frac{1+\bar{x}^2}{x} \ln \frac{p_\perp^2}{M_Z^2}, \tag{4.39}$$

$$F_{W,0}(x) = \frac{g^2}{4\pi} \frac{\bar{x}}{x}, \tag{4.40}$$

$$F_{Z,0}(x) = \frac{g^2}{4\pi c_w^2} \left[(t_3^f - 2q_f s_w^2)^2 + (t_3^f)^2 \right] \frac{\bar{x}}{x}. \tag{4.41}$$

Here, t_3^f is the $SU(2)_L$ quantum number of the fermion ($\pm\frac{1}{2}$) and q_f is the particle charge divided by the positron charge.

The longitudinal structure functions do not rise at high transverse momenta, so they do not need a transverse momentum cutoff. The reason is that, except for the forward direction, longitudinal vector boson emission can be approximated by Goldstone emission which is suppressed by the small fermion mass. The emission of transverse vector bosons extends to higher transverse momenta, and for them the emission probability logarithmically depends on the cutoff $p_\perp$.

Finally, the derivation is repeated for the second intermediate vector boson to end up with the formula

$$\sigma(s) = \sum_{\lambda_1 \lambda_2} \int dx_1 \, dx_2 \, F_{\lambda_1}(x_1) \, F_{\lambda_2}(x_2) \, \hat{\sigma}_{\lambda_1 \lambda_2}(x_1 x_2 s) \tag{4.42}$$

for vector boson fusion processes, where $\hat{\sigma}$ is the on-shell cross section of the subprocess. Analogous formulae hold for differential cross sections.

As an example, for the production of a heavy Higgs boson the subprocess cross section is

$$\hat{\sigma}(s) = 2\pi \delta(x_1 x_2 s - M_H^2) \, \Gamma(H \to W_L W_L) \tag{4.43}$$

with the partial Higgs width [cf. (3.68)]

$$\Gamma(H \to W_L W_L) = \frac{\lambda}{8\pi} M_H = \frac{M_H^3}{16\pi v^2}. \tag{4.44}$$

The resulting total cross section is [73]

$$\sigma(s) = \frac{\alpha^3}{16 M_W^2 s_w^6} \left[\left(1 + \frac{M_H^2}{s} \right) \ln \frac{s}{M_H^2} - 2 + 2\frac{M_H^2}{s} \right]. \tag{4.45}$$

As expected, this cross section increases logarithmically with energy.

While the EWA is valuable for estimating the cross section for particle production from Goldstone scattering in the region of very high invariant mass, in the sub-TeV region it is less reliable. In many cases, the condition that the subprocess cross section is slowly varying on the scale of M_W is not satisfied. For instance, Higgs pair production in WW scattering is of interest mainly for Higgs masses not much larger than M_W. In this case, the EWA can be off by more than a factor of two. Moreover, in the EWA the transverse momentum of the WW system is neglected. As we will discuss below, an accurate knowledge of the $p_\perp$ distribution is essential in isolating the signal from the background, so the EWA is not very useful for this application.

Improving the EWA is possible in some respects, but there are principal problems. To be specific, consider quasielastic vector boson scattering. In the spirit of the approximation, it is consistent to combine the EWA with the equivalence theorem (Section 3.1.1) and estimate total cross sections from Goldstone scattering and longitudinal structure functions. Both approximations are simultaneously valid if $s \gg M_W^2$, $M(WW) \gg M_W$, and if the subprocess scattering occurs at large angle. To go beyond this approximation, one has to take transversal gauge bosons into account [48]. (Note that the transverse structure functions are logarithmically enhanced, and there are more transversal than longitudinal degrees of freedom, so the corrections are larger than one might naively think.) Transversal gauge bosons can also be emitted from fermion lines, which introduces Feynman diagrams which do not have the fusion topology. Due to gauge invariance they cannot be separated from the subset of fusion diagrams. Implementing the EWA for the fusion diagrams without dropping the non-fusion part spoils delicate gauge cancellations. As a consequence, the only consistent choice is either to take the effect of longitudinal gauge bosons only, in which case one may also apply the Equivalence Theorem, or to do a calculation without approximations which takes all degrees of freedom into account.

4.3 Lepton Colliders

In the same spirit that the experiments at LEP1, LEP2, and SLC have been carried out, by measurements at future lepton colliders one will increase the accuracy in the determination of electroweak precision observables. Integrated luminosities in the ab^{-1} range by themselves lead to an improvement of more than an order of magnitude in the overall statistical error for similar energies. Moreover, the collider energy of the planned projects (NLC, JLC, TESLA) is assumed to be considerably higher than in previous colliders, about 500 GeV initially with the prospect to increase this value to 800 GeV . . . 1000 GeV.

Probing the electroweak symmetry-breaking sector in leptonic collisions has a number of advantages. The clean environment and the possibility of checking the momentum balance both in longitudinal and transverse directions (with some smearing due to initial-state radiation, beam energy spread and beamstrahlung) makes it possible to use essentially all final states for the analysis. In the case of W and Z bosons, only a lesser fraction of the decays contains charged leptons which are easy to identify at any collider. In leptonic collisions, momentum balance makes neutrino final states also accessible by deriving the missing momentum and invariant mass. Finally, the quark branching ratios dominate the vector boson decay modes. One can make use of them if a sufficient energy resolution can be achieved such that, for instance, W and Z can be distinguished in the dijet mode. Similar statements hold for top quarks (which decay into W and b) and Higgs bosons (which decay dominantly either into b quarks, or, for larger masses, into W and Z bosons).

Further advantages of lepton colliders include the small energy spread which makes it possible to scan thresholds, and the possibility of polarizing the incoming particles. The disadvantage, of course, is the limited energy reach compared to hadron colliders, at least with the current technology. This can only be compensated by precision, which calls for high luminosity, so that if new physics cannot be probed directly, it has to be deduced from precisely measuring the parameters of the low-energy effective theory.

4.3.1 Associated Production

As discussed in Section 4.2, if the available energy is limited, the amplitudes which are of interest for EWSB can be probed by processes of the type

$$e^+e^- \to ZXX \tag{4.46}$$

4.3.1.1 Resonance Production

The natural observable to check is the invariant mass of the XX state. If one does an inclusive measurement, any heavy particle coupled to ZZ (and $Z\gamma$) will manifest itself as a resonance in the recoil distribution. The prime candidate is, of course, a Higgs boson whose properties can be studied in detail. Using the Z decay channels which are simplest to analyze (e.g., $Z \to \mu^+\mu^-$) and correcting for initial-state radiation and beamstrahlung, the XX invariant mass can be calculated from the beam energy and the Z energy and momentum. In this way, one can derive the resonance mass, its width if it is not too narrow, and the absolute value of the coupling to Z bosons.[3] In addition, the decay branching ratios are directly obtained from the analysis of the exclusive contributions to XX. Finally, the angular distribution of the Z boson is a probe of the resonance spin.

The analysis performed for the TESLA study shows that with realistic assumptions about beam spectra and detector performance, one will be able to measure the total cross section of a light Higgs resonance ($120\ldots160\,\mathrm{GeV}$) with a statistical uncertainty of less than 3 %, assuming an integrated luminosity of $\int \mathcal{L} = 500\,\mathrm{fb}^{-1}$ at $\sqrt{s} = 350\,\mathrm{GeV}$ [75]. Furthermore, branching ratios for the decay of such a resonance can be measured with the relative accuracies given in the following table [76]:

---

[3] One has to correct for double counting if the state XX itself contains a Z boson.

	$M_H = 120\,\mathrm{GeV}$	$M_H = 140\,\mathrm{GeV}$	$M_H = 160\,\mathrm{GeV}$
$H \to b\bar{b}$	2.4 %	2.6 %	6.5 %
$H \to c\bar{c}$	8.3 %	19.0 %	
$H \to gg$	5.5 %	14.0 %	
$H \to \tau^+\tau^-$	5.0 %	8.0 %	

4.3.1.2 Higgs Pairs

In Section 3.3.1 we argued that for a measurement of the Higgs $SU(2)_L$ quantum numbers and of the trilinear term in the Higgs potential one needs the observation of Higgs pair production [77, 78, 27]. If the available energy is limited, at an e^+e^- collider this is only possible via the double-Higgsstrahlung process

$$e^+e^- \to ZHH(\to f\bar{f}b\bar{b}b\bar{b}), \tag{4.47}$$

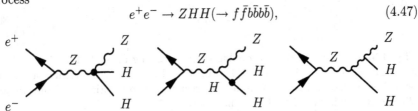

where in the low-mass range the Higgs predominantly decays into $b\bar{b}$. If the invariant mass distribution of the H pair could be measured in some detail, it would be possible to fit both parameters, the $HHZZ$ gauge coupling and the HHH trilinear vertex (3.115, 3.116). Unfortunately, the cross section for this process is so small that it is barely possible to detect the signal at all. Even with optimistic assumptions for the achievable collider performance (e.g., $1\,\mathrm{ab}^{-1}$ of integrated luminosity with electron and positron polarization at $500\,\mathrm{GeV}$ c.m. energy), there will be not more than a few hundred signal events, cf. Fig. 4.3.

This number does not include the detection efficiency, or the branching ratios for the Z and H decays. Fortunately, the background for this kind of process (four b-quarks in the final state) is manageable. Simulation studies show that, if the $HHZZ$ coupling is fixed at the SM value, a statistical accuracy of better than 20 % in the measurement of the trilinear Higgs coupling ($M_H = 120\,\mathrm{GeV}$, $\sqrt{s} = 500\,\mathrm{GeV}$, $\int \mathcal{L} = 1\,\mathrm{ab}^{-1}$) can be expected [79]. Clearly, a two-parameter fit of both couplings will have an error that is significantly bigger.

The phenomenology of this class of processes becomes much richer if there is an extended Higgs sector. For instance, in supersymmetric models there are five physical Higgs scalars, and six trilinear scalar couplings have to be determined [78]. In the presence of Higgs singlets, the situation becomes even more complicated. On the other hand, there are many more final states (including associated production where the Z boson is replaced by a pseudoscalar Higgs

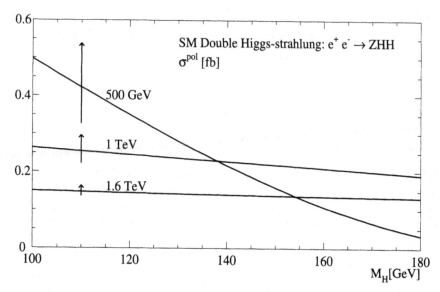

Fig. 4.3. The cross section for double Higgs-strahlung in the SM at three collider energies: 500 GeV, 1 TeV, and 1.6 TeV. The electron/positron beams are taken oppositely polarized. The vertical arrows correspond to a variation of the trilinear Higgs coupling from 1/2 to 3/2 of the SM value [78]

A), and resonance production increases the cross section by orders of magnitude in some regions of parameter space, making a measurement of a subset of the parameters fairly easy (see, e.g., Fig. 4.4).

Triple Higgs production is the only process where the quartic Higgs coupling can be accessed. Unfortunately, the cross section for this process is so small that it is probably inaccessible for colliders in the foreseeable future. Fig. 4.5 summarizes the total cross sections for single and multiple Higgs production in various processes in the TeV energy range.

4.3.2 WW Fusion

4.3.2.1 Higgs Bosons

For sufficiently high energy, the cross section of the WW fusion process

$$e^+ e^- \to \bar{\nu}_e \nu_e H(\to \bar{\nu}_e \nu_e b \bar{b}, \ \bar{\nu}_e \nu_e W^+ W^-, \ldots) \tag{4.48}$$

becomes significant and eventually overtakes the Higgs-strahlung cross section. For a heavy Higgs boson, it is approximately given by (4.45), while for a light Higgs boson there are substantial deviations from this formula. Since the Z boson in Higgs-strahlung can decay into neutrinos, there is an interference between the two processes. However, for sufficiently high collider energy

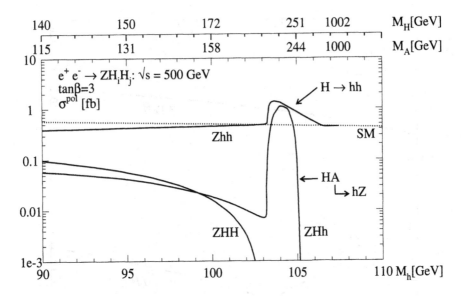

Fig. 4.4. Cross sections for the processes Zhh, ZHh, and ZHH for $\sqrt{s} = 500\,\text{GeV}$ and $\tan\beta = 3$, including mixing effects in the supersymmetric parameters ($A = 1\,\text{TeV}$, $\mu = -1\,\text{TeV}$) [78]

they can easily be kinematically separated: The invisible invariant mass, for instance, peaks at the Z mass in Higgs-strahlung while it tends to be near the kinematical limit in WW fusion. The similar ZZ fusion process is suppressed compared to WW fusion by about a factor of 10; this is a consequence of the small eeZ coupling.

An inclusive measurement of the Higgs production cross section is not possible in WW fusion: The Higgs mass would have to be reconstructed from the recoil of the neutrino pair which is not detected. One would need ZZ fusion for that purpose, but due to the small cross section one probably cannot improve the Higgs-strahlung measurement in that way. However, one can measure the production rate in exclusive channels, which is proportional to the $H \to WW$ partial decay width multiplied by the respective branching ratios of the Higgs boson. Since the branching ratios are independently known from Higgs-strahlung (making use of the inclusive measurement in Higgs-strahlung), the $H \to WW$ partial decay width can be extracted. Combining this with the measured value of the $H \to WW$ branching ratio, one can determine the total Higgs decay width in the low Higgs mass range where it is not resolvable by the detector. Studies for the TESLA collider parameters show that an accuracy from 6 % ($M_H = 120\,\text{GeV}$) to 13 % ($M_H = 160\,\text{GeV}$) can be achieved in this way [81]. This allows, for instance, for a limit to be set on the branching fraction of invisible Higgs decays just as it has been done at LEP 1 for the Z boson.

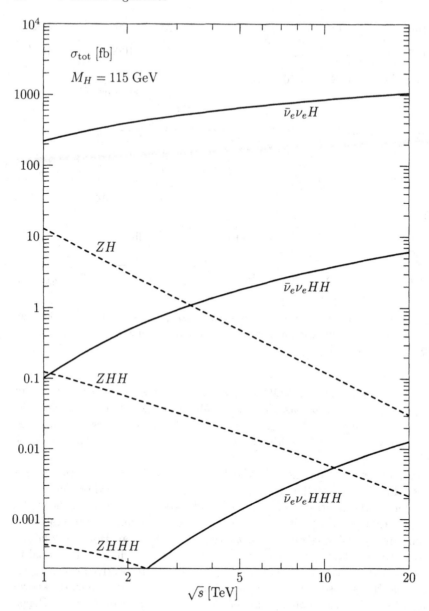

Fig. 4.5. Total cross sections for single and multiple SM Higgs production in unpolarized e^+e^- collisions ($M_H = 115\,\text{GeV}$). The numbers were obtained using WHIZARD [72]

Higgs pair production in WW fusion is also of interest [80, 78]:

$$e^+e^- \to \bar{\nu}_e\nu_e HH \qquad\qquad (4.49)$$

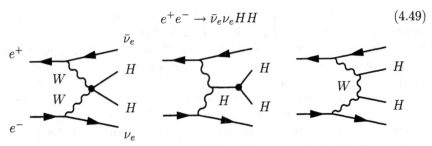

The cross section is small for low collider energies where the Higgs-strahlung cross sections peak, but it rapidly increases for higher energies (Figs. 4.5, 4.6). The variation of the cross section with the trilinear coupling is in the opposite direction compared with Higgs-strahlung.

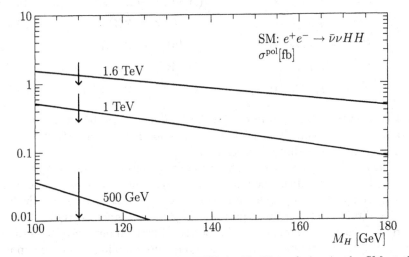

Fig. 4.6. The total cross section for WW double-Higgs fusion in the SM at three collider energies: 500 GeV, 1 TeV, and 1.6 TeV. The vertical arrows correspond to a variation of the trilinear Higgs coupling from 1/2 to 3/2 of the SM value [78]

If the Higgs mass is large enough, it turns into a broad resonance in WW scattering. This case has been discussed at length in Section 3.2.3. Here we just note that for a full calculation, one has to include the transversal degrees of freedom which considerably complicates the problem of describing the resonance shape beyond the leading order. This has not yet been done. If this scenario was realized in nature, it would be important to accurately measure the resonance shape and, in particular, its high-energy tail to get information about the breakdown of the Standard Model at higher energies.

4.3.2.2 Quasielastic Scattering

Apart from the Minimal Standard Model, the case which is most straightforward to investigate is the Higgs-less scenario where there is, below a certain
energy threshold, no direct signature of new physics. In that case, apart from
the precision observables considered in Section 4.1, the anomalous quartic couplings $\mathcal{L}_{4,5,6,7,10}$ of (longitudinally polarized) W and Z bosons are the prime
focus [82]. While we have considered the on-shell quasielastic scattering amplitudes in Section 3.1, in practice they are embedded in fermionic processes at
colliders, and one has to check the sensitivity to the couplings in the presence
of reducible and irreducible backgrounds.

Leaving aside rescattering in WW pair production (see Section 4.2.1), the
processes of main interest at electron-positron colliders and the scattering
amplitudes probed by them are [83]:

$$e^+e^- \to W^+W^-Z \qquad\qquad (W^+W^- \to ZZ), \qquad\qquad (4.50)$$

$$e^+e^- \to ZZZ \qquad\qquad (ZZ \to ZZ), \qquad\qquad (4.51)$$

$$e^+e^- \to \bar{\nu}_e\nu_e W^+W^- \qquad\qquad (W^+W^- \to W^+W^-), \qquad\qquad (4.52)$$

$$e^+e^- \to \bar{\nu}_e\nu_e ZZ \qquad\qquad (W^+W^- \to ZZ), \qquad\qquad (4.53)$$

$$e^-e^- \to \nu_e\nu_e W^-W^- \qquad\qquad (W^-W^- \to W^-W^-). \qquad\qquad (4.54)$$

The last process, calling for an e^-e^- collider, does not probe an independent
interaction but it uniquely projects the scattering matrix onto the $I = J = 2$
channel which is thus directly accessible [84].

In the low-energy range, only the triple vector boson production processes
(4.50, 4.51) are accessible. The impact of the anomalous four-boson couplings
on them is only visible if they are unexpectedly large, but they provide the first
available direct limits on those parameters [85]. At higher energies, they interfere with vector boson fusion (all processes described here have six-fermion
final states), and a combined fit to the anomalous couplings would be appropriate. However, since the kinematics of triple vector boson production and
vector boson fusion are quite distinct at sufficiently high energies, they are
usually treated separately.

In addition, the following processes are important:

$$e^+e^- \to e^\pm\nu_e W^\mp Z \qquad\qquad (W^+W^- \to ZZ), \qquad\qquad (4.55)$$

$$e^+e^- \to e^+e^- W^+W^- \qquad\qquad (W^+W^- \to ZZ), \qquad\qquad (4.56)$$

$$e^+e^- \to e^+e^- ZZ \qquad\qquad (ZZ \to ZZ), \qquad\qquad (4.57)$$

together with their conterparts in e^-e^- collisions. The signal processes contained here have lower cross sections (due to the small Zee coupling) and
larger background, so their value is limited. Nevertheless, the three independent amplitudes of quasielastic vector boson scattering can be probed in
various different channels, and exploiting angular distributions it should be

possible to extract limits for all five anomalous couplings $\alpha_{4,5,6,7,10}$. To simplify the discussion, for obtaining the numerical results shown below it has been assumed that the custodial symmetry $SU(2)_C$ is conserved, such that the $SU(2)_C$-violating couplings $\alpha_{6,7,10}$ are suppressed and can be ignored in a first approximation.

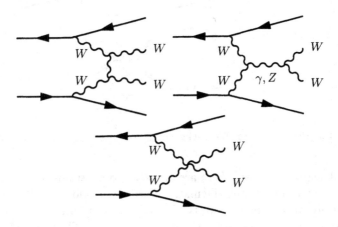

Fig. 4.7. Diagrams contributing to the strong WW scattering signal

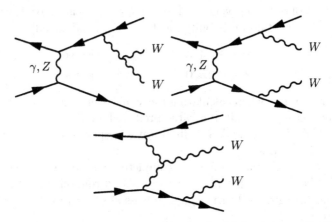

Fig. 4.8. Typical diagrams contributing to the irreducible background for the strong WW scattering signal

The (quasi-)elastic WW scattering signal corresponds to the generic diagrams depicted in Fig. 4.7. However, there are also Feynman diagrams contributing to (4.52–4.54) which do not contain WW scattering as a subprocess (cf. Fig. 4.8). This irreducible background is not negligible and must be taken into account in the analysis.

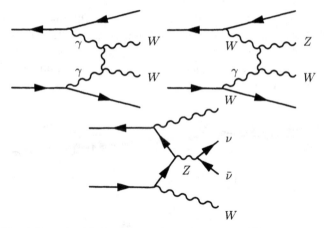

Fig. 4.9. Partially reducible backgrounds to the strong WW scattering signal

In all signal processes there are already two neutrinos present in the final state, therefore important kinematic information is lost if a W boson decays leptonically (or a Z boson into two neutrinos). In particular, the c.m. energy of the subprocess cannot be determined in that case. Therefore, the analysis is easiest for hadronic W, Z decays and for decays of the Z boson into electrons and muons. Furthermore, an error in the dijet invariant mass is introduced by the limited energy resolution of the calorimeters, which leads to the rejection of a fraction of di-boson events and to the misidentification of W vs. Z bosons. This misidentification probability has been estimated as [83]

$$W \to 85\% \ W, \ 10\% \ Z, \ 5\% \ \text{rejected}, \tag{4.58}$$

$$Z \to 22\% \ W, \ 74\% \ Z, \ 4\% \ \text{rejected}. \tag{4.59}$$

Thus, for example, when calculating the signal event rate in the ZZ detection mode, one has to include the rates predicted by 55%, 7%, and 1% of the partonic ZZ, $W^\pm Z$ and W^+W^- final states, respectively, to account for final-state misidentification.

Since the final state cannot be completely resolved experimentally in all cases, further background processes will play a role (cf. Fig. 4.9). The most important background to the signal process $e^+e^- \to \bar{\nu}\nu W^+W^-$ is generated by the reaction

$$e^+e^- \to W^+W^-e^+e^-, \tag{4.60}$$

which is built up primarily by the subprocess $\gamma\gamma \to W^+W^-$. In this process most of the electrons and positrons are emitted in forward direction so that they cannot be detected. A similar background is introduced by the misiden-tification of vector bosons in jet decays:

$$e^+e^- \to W^\pm Z e^\mp \nu. \tag{4.61}$$

An irreducible background is also generated by three-boson final states,

$$e^+e^- \rightarrow W^+W^-Z, \tag{4.62}$$

with the Z decaying into neutrino pairs. Similar backgrounds (less dangerous for the ZZ final state) exist for the other processes.

The total cross sections for the signal and background processes at various e^+e^- collider energies, calculated for the reference point $\alpha_4 = \alpha_5 = 0$, are summarized in Table 4.1.

Process	800 GeV	1.6 TeV	Factor
$W^+W^-\bar\nu\nu$	11	56	1
$W^+W^-e^+e^-$	628	1979	1
$W^\pm Ze^\mp\nu$	39	173	0.26
$W^+W^-(Z \rightarrow \bar\nu\nu)$	13	11	1
$ZZ\bar\nu\nu$	4	26	1
ZZe^+e^-	2	4	1
$W^\pm Ze^\mp\nu$	39	173	0.13
$W^+W^-e^+e^-$	628	1979	0.018
$ZZ(Z \rightarrow \bar\nu\nu)$	0.6	0.4	1
$W^-W^-\nu\nu$	14	67	1

Table 4.1. Total cross sections in fb for various processes. Dectection efficiencies and branching ratios are not included. Including final-state misidentification, the numbers should be multiplied by the relative weighting factor given in the last column which accounts for final-state misidentification in the corresponding detection mode (W^+W^-, ZZ, or W^-W^-) [82]

Background reduction is essential for isolating the strong scattering signal. Analyzing the kinematics, one can derive the following strategy:

1. Requiring a minimal invisible invariant mass of the order $M_{\text{inv}}(\bar\nu\nu) > 150 \ldots 200$ GeV removes the events with neutrinos from Z decay together with backgrounds from W^+W^- and QCD four-jet production. The signal is not affected.

2. Selecting central events $[|\cos\theta(W/Z)| \lesssim 0.8]$ with considerable W, Z transverse momentum ($p_\perp(W/Z) \gtrsim 150$ GeV) removes events dominated by t-channel exchange in the subprocess.

3. The background from $\gamma\gamma$ fusion is drastically reduced if a veto for forward-going electrons and positrons is applied and, at the same time, a minimum $p_\perp$ of the vector boson pair, equivalent to the fermion $p_\perp$, is required. With the current design of a future e^+e^- collider detector, the angular coverage in the forward direction is sufficient to make this possible without losing too much of the signal.

4. Since the impact of the anomalous couplings $\mathcal{L}_4$ and $\mathcal{L}_5$ increases with the energy of the subprocess, one should not use events with low invariant WW or ZZ mass.

5. Exploiting the angular distribution of the fermions in the W and Z decays allows for projecting onto longitudinal polarization states.
6. Polarizing the incident electron beam makes it possible to double the signal cross section while the γ-induced background is unaffected. The signal-to-background ratio is improved even more if positron polarization is available as well.

A cut-based analysis following this strategy yields the sensitivity estimate shown in Fig. 4.10, an exclusion region around the assumed reference value $\alpha_4 = \alpha_5 = 0$. The plot consists of two ellipse-shaped regions which intersect in two regions, one of them in the center of the plot, the other one outside of the visible area.

The latter region can be excluded by a more elaborate analysis. For the TESLA parameters, such a study has been performed, using a full six-fermion simulation of the signal and background [72] and including all interference terms and angular correlations of the final-state fermions [86]. From a likelihood analysis of simulated event samples, accounting for detector performance, one derives the exclusion countours of Fig. 4.11. Comparing Figs. 4.10 (top) and 4.11 (note the different normalization), the allowed regions are similar in area and shape.

4.3.2.3 Other Final States

Beyond quasielastic scattering one would like to be able to observe top and bottom quarks and, if they exist, new strongly-coupled states such as technipions in WW scattering processes. The rise of the cross section depends very much on the pair production threshold, such that for realistic next-generation collider energies $\bar{t}t$ production from WW scattering is probably observable in the resonant case only [87]. A similar statement holds for other particle pairs if their mass is comparable. The situation changes if one has a high-luminosity lepton collider in the TeV range (e^+e^- or $\mu^+\mu^-$).

The principal strategy for isolating the signal is similar to the quasielastic processes considered above, with the exception that one would expect heavy quarks to accompany the produced (longitudinal) vector bosons. In the multi-TeV range one could even imagine the possibility of producing pairs of new constituents (e.g., techniquarks) which "hadronize" into broad jets of W_L, Z_L, technipions and heavy quarks.

4.4 Hadron Colliders

At hadron colliders, the class of processes which are related to electroweak symmetry breaking is quite similar to those at lepton colliders. However, while signals are similar, with leptons replaced by quarks, the big difference lies in the amount of background that has to be overcome. The large rates of

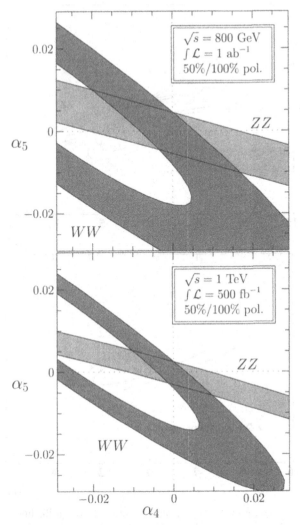

Fig. 4.10. Allowed regions in the α_4-α_5 plane resulting from a cut-based analysis of quasielastic vector boson scattering in polarized e^+e^- collisions for two different sets of collider parameters. The exclusion contours are based on the hypothesis that the actual values are $\alpha_4 = \alpha_5 = 0$ [82]

hadronic background events can be beaten by concentrating on leptonic final states. This removes a major part of the available cross section. Furthermore, no information about the longitudinal momenta of the incoming particles is available, so invariant mass reconstructions are difficult. On the other hand, the accessible energy is typically higher than for lepton colliders, so precision measurements can be replaced by direct observation.

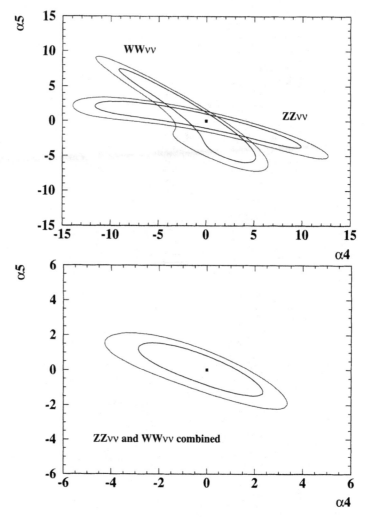

Fig. 4.11. Allowed regions in the α_4-α_5 plane resulting from a likelihood analysis of simulated six-fermion event samples corresponding to quasielastic vector boson scattering in polarized e^+e^- collisions [86]. The exclusion contour is based on the hypothesis that the actual values are $\alpha_4 = \alpha_5 = 0$. The assumed integrated luminosity is $\int \mathcal{L} = 1\,\mathrm{ab}^{-1}$ at an energy of $\sqrt{s} = 800\,\mathrm{GeV}$ with 80% (40%) electron (positron) polarization. The inner and outer contours represent 68% C.L. and 90% C.L. limits, respectively. Note that a factor of $16\pi^2$ is absorbed in the definition of α_4 and α_5; to compare the results with Fig. 4.10, for instance, the numbers on the axes have to be multiplied by $1/16\pi^2 = 0.00633$

4.4.1 Gluon Fusion

At high-energy much of the momentum of protons is carried by gluons. This makes the production of colored particles, which can proceed via gluon-gluon fusion, the dominant class of processes. Unfortunately, since the QCD gauge group does not appear to be directly involved in electroweak symmetry breaking (an assumption that may be wrong), states associated with electroweak symmetry breaking are not generically expected to have color. Thus, the production of particles associated with electroweak symmetry breaking in gluon-gluon fusion involves higher-order diagrams.

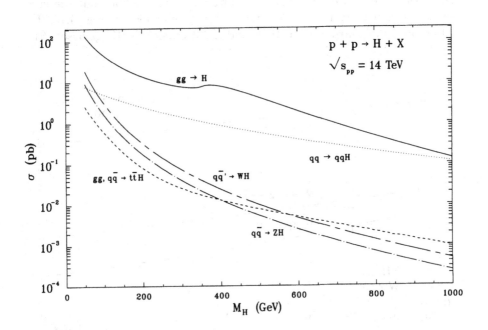

Fig. 4.12. Total cross section for single SM Higgs production at LHC [88]

4.4.1.1 Higgs Production

A Higgs boson couples to gluons via a top-quark loop. The cross section for this process is large compared to electroweak processes (Fig. 4.12) [88], and if the Higgs decay products could be detected, precision Higgs physics would be possible. Unfortunately, for a light Higgs boson, the detection of the main decay channel $H \rightarrow b\bar{b}$ is very difficult in a hadronic environment, even the very rare $\gamma\gamma$ decay mode looks more promising. For higher Higgs masses, although the production cross section drops, the WW and ZZ channels with leptonic final states make the detection easier. If different decay modes can be observed, a number of ratios of branching ratios can be extracted. However, without the possibility for an inclusive measurement (such as $e^+e^- \to ZX$), the normalization (i.e., the total Higgs width) cannot be unambiguously fixed unless it is large enough to be measured directly [89].

Higgs pairs can be produced in gluon fusion as well [90]:

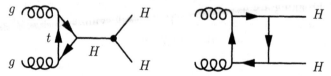

Here, the detection of the $b\bar{b}b\bar{b}$ final state might be possible, but the cross section is much lower than for single Higgs production, cf. Fig. 4.13. The situation is more favorable for WW and ZZ decays of the Higgs boson [91].

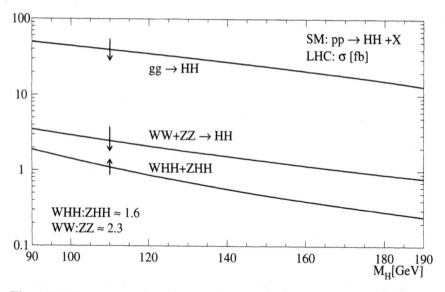

Fig. 4.13. The cross sections for gluon fusion, WW/ZZ fusion, and double Higgs-strahlung WHH, ZHH in the SM. The vertical arrows correspond to a variation of the trilinear Higgs coupling from 1/2 to 3/2 of the SM value [91]

4.4.1.2 W Pairs

Looking at the Feynman diagrams, one can interpret the processes $gg \to W^+W^-$ and $gg \to ZZ$ [92] as probes of the $WW \to t\bar{t}$ and $ZZ \to t\bar{t}$ amplitudes:

The value of the loop diagrams is related to these on-shell amplitudes by a dispersion relation. In the presence of a Higgs boson, this implies that the top Yukawa coupling can be measured (if the overall normalization is known) even if the decay channel $H \to t\bar{t}$ is closed. This is true only if there are no other colored states strongly coupled to the Higgs boson. If an independent measure of the top Yukawa coupling can be found, e.g., $t\bar{t}H$ production, the existence of such states can be detected by comparing the two cross sections. However, since W/Z pair production in gluon fusion cannot be distinguished from W and Z pair production in quark annihilation which occurs at tree level, there is an inherent uncertainty in this extraction.

4.4.2 Associated Higgs Production

Compared to gluon fusion, the cross section for Higgs-strahlung is small in high-energy hadronic collisions ([88], cf. Fig. 4.12). However, the recoiling Z which can decay, e.g., into $\mu^+\mu^-$ makes the Higgs detection easier, and the main $b\bar{b}$ decay channel of a light Higgs boson becomes easier to access. As opposed to the situation at a lepton collider with well-defined beam energy, the absence of partonic longitudinal momentum information in hadronic collisions makes an inclusive measurement of ZX final states impossible. Nevertheless, observing this channel adds important information. Finally, the cross section for double Higgs-strahlung is very small (Fig. 4.13) and may be impossible to extract from the background [93, 91].

4.4.3 WW and ZZ Fusion

In the absence of a Higgs boson, the cross section for WW and ZZ quasielastic scattering is large enough to be detectable at the LHC even if hadronic decay modes have to be dropped. The signature of these processes [94],

$$q_1 q_2 \to q_3 q_4 W^+ W^-, \qquad (4.63)$$

$$q_1 q_2 \to q_3 q_4 W^\pm Z, \qquad (4.64)$$

$$q_1 q_2 \to q_3 q_4 ZZ, \qquad (4.65)$$

$$q_1 q_2 \to q_3 q_4 W^\pm W^\pm, \qquad (4.66)$$

are two energetic forward jets together with (for leptonic W, Z decays) no jet activity in the central region. A forward jet tag and a central jet veto are very helpful in reducing the background.

As the list of processes shows, all isospin and partial-wave amplitudes are accessible in principle, although the different branching ratios and kinematic properties of the final state make the detection efficiencies quite different. Note that there are contributions both from WW and from ZZ fusion which cannot be separated. The easiest final-state channels appear to be $W^\pm Z$ (only one neutrino) and W^-W^-/W^+W^+ (like-sign leptons) while W^+W^- (two neutrinos) and ZZ (low branching ratio for the four-lepton state) are more difficult. This makes the measurements at lepton and hadron colliders complementary to each other.

The observation of the processes (4.63–4.66) is crucial if no light Higgs boson can be detected at a hadron collider. If the Higgs boson couplings deviate significantly from their SM values, at LHC a light Higgs state could escape detection in the rare $\gamma\gamma$ channel. However, the existence or absence of a scalar state which unitarizes $W_L W_L$ scattering amplitudes at high energies can be decided by measuring them directly at LHC or at a next-generation e^+e^- collider.

Beyond the question of weak or strong coupling in the quasielastic scattering amplitudes, one would like to extract limits or values for the anomalous couplings which enter at next-to-leading order in the low-energy expansion. For LHC, cut-based sensitivity studies have been performed in the same spirit as for lepton colliders [95]. Combining the information from all four channels, the sensitivity turns out to be slightly worse than that achievable at a high-luminosity lepton collider with 800 GeV c.m. energy (Figs. 4.10, 4.11).

The high energies achievable in vector boson scattering at hadron colliders make it conceivable that one is not just sensitive to the NLO anomalous couplings in the chiral Lagrangian, but that one can actually produce new resonances strongly coupled to Goldstone bosons. A detailed survey of a number of resonant and non-resonant models (cf. Chapter 3) and the prospects of observing the corresponding signatures in an actual experiment was performed for the SSC and LHC collider projects [94]. The analysis shows that within the kinematical reach of the collider a clear separation of models with distinctive features (e.g., a scalar or a vector resonance) is in fact possible.

5

Outlook

Despite the great success the Standard Model has achieved in recent years in describing electroweak interactions, in terms of the questions posed in the introduction we have not made significant progress. Reading texts written twenty years ago, one observes that the arguments favoring or disfavoring scenarios of electroweak symmetry breaking have not changed very much, nor have we come much closer to a solution of the hierarchy problem.

Triggered both by precision data and by theoretical considerations, there is a slight preference for supersymmetric scenarios which accomodate electroweak symmetry breaking in conjunction with the unification of electroweak and strong interactions somewhat below the Planck scale [35]. However, the attempts to find a compelling model which naturally generates supersymmetry breaking, thus solving the hierarchy problem, have remained unsuccessful so far. On the other hand, strong-interaction models do not fare much better: they have to deal with the complicated pattern of fermion masses and mixings which makes realistic models awkward to construct [36]. The recent development of extra-dimension scenarios [96, 31] has contributed even more to the present confusion: While providing an elegant alternative approach to the hierarchy problem, their predictions for low-energy observables are rather uncertain, such that it is impossible to judge their validity without having experimental data from a new generation of colliders at hand.

Fortunately, if the collider projects that are currently being planned can be realized, our knowledge about these issues will significantly improve. As the effective-Lagrangian approach proves, the present theory of electroweak interactions either fails in the TeV energy range, giving way to a different, more fundamental setup, or one will observe Higgs scalars which point to a weakly-interacting scenario. In the latter case, if the hierarchy problem has a solution at all, there will be even more new particles beyond the Higgs sector to be discovered. The LHC will shed light on this, hopefully uncovering part of the new spectrum still hidden from us. However, due to the overwhelming hadronic background, LHC experiments have limitations in precisely probing the Higgs sector, while the situation in e^+e^- collisions is much more favorable.

Therefore, the construction of a high-energy, high-luminosity linear collider is crucial for the origin of electroweak symmetry breaking to be uncovered in the foreseeable future.

Ultimately, the energy reach of colliders in the post-LHC era should be sufficient to look beyond the characteristic scale of electroweak symmetry breaking, even if no light Higgs boson exists. One can only speculate about the opportunities that open up there [97]. For instance, in a strongly-interacting scenario one could be able to directly produce pairs of the constituents which make up the Goldstone bosons of electroweak symmetry breaking. Alternatively, in a weakly interacting scenario, if electroweak symmetry breaking (or supersymmetry breaking) is mediated by gauge interactions, one could directly observe the corresponding messenger states. If an extra-dimension scenario turns out to be realized, one might look into details of quantum gravity. In any case, while the physics of electroweak symmetry breaking and mass generation will have become much clearer after the measurements described in this review have been performed, it is conceivable that the origin of these effects cannot be fully uncovered at once, and another generation of colliders in the multi-TeV range will be necessary to proceed further in that direction.

References

1. M.E. Peskin and D.V. Schroeder, *An Introduction to Quantum Field Theory*, Addison-Wesley, 1995; S. Weinberg, *The Quantum Theory of Fields (Vol. 2)*, Cambridge University Press, 1996; H. Georgi, *Weak interactions and modern particle theory*, Addison-Wesley, 1984. J.F. Donoghue, E. Golowich, and B.R. Holstein, *Dynamics of the Standard Model*, Cambridge University Press, 1992; A. Dobado, A. Gómez-Nicola, A.L. Maroto, and J.R. Peláez, *Effective Lagrangians for the Standard Model*, Springer, 1997; J.F. Gunion, H.E. Haber, G. Kane, and S. Dawson, *The Higgs Hunter's Guide*, Addison-Wesley, 1990.
2. S.L. Glashow, Nucl. Phys. 22 (1961) 579; A. Salam, in: N. Svartholm (ed.), *Elementary Particle Theory*, Almqvist and Wiksells, Stockholm, 1968; S. Weinberg, Phys. Rev. Lett. 19 (1967) 1264.
3. G. 't Hooft, Nucl. Phys. B33 (1971) 173; Nucl. Phys. B35 (1971) 167; G. 't Hooft and M. Veltman, Nucl. Phys. B50 (1972) 318.
4. H. Fritzsch and M. Gell-Mann, in: J.D. Jackson and A. Roberts (eds.), *Proc. XVI Int. Conf. on High Energy Physics*, Fermilab, 1972.
5. D. Gross and F. Wilczek, Phys. Rev. Lett. 30 (1973) 1343; H.D. Politzer, Phys. Rev. Lett. 30 (1973) 1346.
6. K. Hagiwara et al. (Particle Data Group), Phys. Rev. D. 66 (2002) 010001; for updates, see: http://pdg.lbl.gov/.
7. L. Susskind, Phys. Rept. 104 (1984) 181.
8. P.W. Higgs, Phys. Rev. Lett. 12 (1964) 132; Phys. Rev. Lett. 13 (1964) 508; Phys. Rev. 145 (1966) 1156; F. Englert and R. Brout, Phys. Rev. Lett. 13 (1964) 321; G.S. Guralnik, C.R. Hagen, and T.W. Kibble, Phys. Rev. Lett. 13 (1964) 585; T.W. Kibble, Phys. Rev. 155 (1967) 1554.
9. S. Weinberg, Phys. Rev. 166 (1968) 1568; S.R. Coleman, J. Wess, and B. Zumino, Phys. Rev. 177 (1969) 2239; C.G. Callan, S.R. Coleman, J. Wess, and B. Zumino, Phys. Rev. 177 (1969) 2247.
10. T. Appelquist and C. Bernard, Phys. Rev. D22 (1980) 200; A. Longhitano, Phys. Rev. D22 (1980) 1166; Nucl. Phys. B188 (1981) 118.
11. S. Weinberg, Physica 96A (1979) 327.
12. For a review, see, e.g.: H. Georgi, Ann. Rev. Nucl. Part. Sci. 43 (1993) 209.
13. This nomenclature originates from: K.G. Wilson, Phys. Rev. B4 (1971) 3174; Phys. Rev. B4 (1971) 3184.

14. N. Cabbibo, Phys. Rev. Lett. 10 (1963) 531; M. Kobayashi and T. Maskawa, Prog. Theor. Phys. 49 (1973) 652.

15. M. Peskin and T. Takeuchi, Phys. Rev. Lett. 65 (1990) 964; Phys. Rev. D46 (1992) 381.

16. H. Georgi, Nucl. Phys. B363 (1991) 301.

17. D. Abbaneo et al. (LEP and SLD Electroweak Working Groups), CERN-EP-2001-098, SLAC-PUB-9498, hep-ex/0112021.

18. J. Gasser and H. Leutwyler, Ann. Phys. (N.Y.) 158 (1984) 142; Nucl. Phys. B250 (1985) 465.

19. C. Burgess and D. London, Phys. Rev. D48 (1993) 4337; C. Burgess, S. Godfrey, H. Konig, D. London, and I. Maksymyk, Phys. Rev. D49 (1994) 6115.

20. R. Haag, *Local Quantum Physics*, Springer, 1992.

21. M. Gell-Mann, P. Ramond and R. Slansky, in: *Supergravity*, eds. P. van Nieuwenhuizen and D. Freedman, North Holland, Amsterdam, 1979;

22. S. Weinberg, Phys. Rev. D13 (1976) 974; Phys. Rev. D19 (1979) 1277; L. Susskind, Phys. Rev. D20 (1979) 2619.

23. R. Haag, Phys. Rev. 112 (1958) 669; H.J. Borchers, Nuovo Cim. 25 (1960) 270; D. Ruelle, Helv. Phys. Acta 35 (1962) 34; J.M. Cornwall, D.N. Levin, and G. Tiktopoulos, Phys. Rev. D10 (1974) 1145, Erratum D11 (1975) 972; M.C. Bergère and Y.-M.P. Lam, Phys. Rev. D13 (1976) 3247; P. Breitenlohner and D. Maison, Commun. Math. Phys. 52 (1977) 11, 39, 55.

24. For reviews, see, e.g.: T. Kugo and I. Ojima, Prog. Theor. Phys. 60 (1978) 1869; Prog. Theor. Phys. 61 (1979) 294; Prog. Theor. Phys. 61 (1979) 644; O. Piguet and S.P. Sorella, *Algebraic renormalization: perturbative renormalization, symmetries and anomalies*, Springer, 1995.

25. D. Ross and M. Veltman, Nucl. Phys. B95 (1975) 135; M. Veltman, Act. Phys. Pol. B8 (1977) 475; Nucl. Phys. B123 (1977) 89; P. Sikivie, L. Susskind, M. Voloshin, and V. Zakharov, Nucl. Phys. B173 (1980) 189.

26. W. Buchmüller and D. Wyler, Nucl. Phys. B268 (1986) 621.

27. V. Barger, T. Han, P. Langacker, B. McElrath, and P.M. Zerwas, MADPH-02-1303, hep-ph/0301097.

28. C. Arzt, M. Einhorn, and J. Wudka, Nucl. Phys. B433 (1995) 41.

29. R. Sekhar Chivukula and N. Evans, Phys. Lett. B464 (1999) 244; Phys. Rev. Lett. 85 (2000) 511.

30. T. Hambye and K. Riesselmann, Phys. Rev. D55 (1997) 7255.

31. L. Randall and R. Sundrum, Phys. Rev. Lett. 83 (1999) 3370; Phys. Rev. Lett. 83 (1999) 4690.

32. G. 't Hooft, in: G. 't Hooft et al. (eds.), *Proc. Cargese Summer Institute* (NATO Advanced Study Institute, Series B: Physics, v. 59), 1979; reprinted in: E. Farhi and R. Jackiw (eds.), *Dynamical gauge symmetry breaking*, World Scientific, 1982; and in: G. 't Hooft (ed.), *Under the spell of the gauge principle*, World Scientific, 1994.

33. N. Arkani-Hamed, A. Cohen, and H. Georgi, Phys. Lett. B513 (2001) 232; N. Arkani-Hamed, A. Cohen, E. Katz, and A. Nelson, JHEP 0207 (2002) 034; N. Arkani-Hamed, A. Cohen, T. Gregoire, and J. Wacker, JHEP 0208 (2002) 020.

34. J. Wess and B. Zumino, Nucl. Phys. B70 (1974) 39;

35. For reviews, see, e.g.: J. Wess and J. Bagger, *Supersymmetry and Supergravity*, Princeton Series in Physics, 1992; P. Fayet and S. Ferrara, Phys. Rept. 32

(1977) 249; H.P. Nilles, Phys. Rept. 110 (1984) 1; R. Barbieri, Riv. Nuovo Cim. 11 (1988) 1; H.E. Haber and G. Kane, Phys. Rept. 117 (1985) 75; N. Polonsky, *Supersymmetry: Structure and Phenomena*, Springer, 2001.

36. For reviews of models of dynamical symmetry breaking, see: R.K. Kaul, Rev. Mod. Phys. 55 (1983) 449; R. Sekhar Chivukula, A. Cohen, and K. Lane, Nucl. Phys. B343 (1990) 554; S.F. King, Rept. Prog. Phys. 58 (1995) 263; G. Cvetic, Rev. Mod. Phys. 71 (1999) 513; C.T. Hill and E.H. Simmons, *Strong Dynamics and Electroweak Symmetry Breaking*, hep-ph/0203079, to appear in Phys. Rept.

37. W. Bardeen, C.T. Hill, and M. Lindner, Phys. Rev. D41 (1990) 1647; C.T. Hill, Phys. Lett. B266 (1991) 419.

38. C.T. Hill, Phys. Lett. B345 (1995) 483; K. Lane and E. Eichten, Phys. Lett. B352 (1995) 382.

39. S.P. Martin, Phys. Rev. D44 (1991) 2892; C.T. Hill, M.A. Luty, and E.A. Paschos, Phys. Rev. D43 (1991) 3011; E. Akhmedov, M. Lindner, E. Schnapka, and J.W.F. Valle, Phys. Rev. D53 (1996) 2752; Phys. Lett. B368 (1996) 270; S. Antusch, J. Kersten, M. Lindner, and M. Ratz, TUM-HEP-491-02, hep-ph/0211385.

40. S. Dimopoulos and L. Susskind, Nucl. Phys. B155 (1979) 237; E. Eichten and K. Lane, Phys. Lett. B90 (1980) 125.

41. J. Goldstone, Nuovo Cim. 9 (1961) 154; Y. Nambu, Phys. Rev. Lett. 4 (1960) 380; J. Goldstone, A. Salam, and S. Weinberg, Phys. Rev. 127 (1962) 965.

42. C.E. Vayonakis, Lett. Nuovo Cim. 17 (1976) 383; M.S. Chanowitz and M.K. Gaillard, Nucl. Phys. B261 (1985) 379; G.J. Gounaris, R. Kögerler, and H. Neufeld, Phys. Rev. D34 (1986) 3257; Y.-P. Yao and C.-P. Yuan, Phys. Rev. D38 (1988) 2237; J. Bagger and C. Schmidt, Phys. Rev. D34 (1990) 264; H.-J. He, Y.-P. Kuang, and X. Li, Phys. Rev. Lett. 69 (1992) 2619; Phys. Rev. D49 (1994) 4842; Phys. Lett. B329 (1994) 278; H.-J. He, Y.-P. Kuang, and C.-P. Yuan, Phys. Rev. D51 (1995) 6463; H.-J. He and W.B. Kilgore, Phys. Rev. D55 (1997) 1515.

43. S. Weinberg, Phys. Rev. Lett. 17 (1966) 616; M.S. Chanowitz, M. Golden, and H. Georgi, Phys. Rev. D36 (1987) 1490.

44. S. Dawson and S. Willenbrock, Phys. Rev. 40 (1989) 2880.

45. O. Cheyette and M. Gaillard, Phys. Lett. B197 (1987) 205.

46. B. Lee, C. Quigg and H. Thacker, Phys. Rev. Lett. 38 (1977) 883; Phys. Rev. D16 (1977) 1519; D. Dicus and V. Mathur, Phys. Rev. D7 (1973) 3111.

47. S.N. Gupta, *Quantum Electrodynamics*, Gordon and Breach, 1981.

48. M.S. Chanowitz, Phys. Rept. 320 (1999) 139.

49. A. Dobado and J.R. Pelaez, Phys. Rev. D56 (1997) 3057.

50. R. Casalbuoni, S. De Curtis, D. Dominici, and R. Gatto, Phys. Lett. B155 (1985) 95; Nucl. Phys. B282 (1987) 235.

51. W. Kilian and K. Riesselmann, Phys. Rev. D58 (1998) 053004.

52. M.H. Seymour, Phys. Lett. B354 (1995) 409.

53. S. Dawson and S. Willenbrock, Phys. Rev. Lett. 62 (1989) 1232; W. Marciano, G. Valencia, and S. Willenbrock, Phys. Rev. D40 (1989) 1725; L. Durand, J.M. Johnson, and J.L. Lopez, Phys. Rev. Lett. 64 (1990) 1215; Phys. Rev. D45 (1992) 3112; K. Riesselmann, Phys. Rev. D53 (1996) 6226; U. Nierste and K. Riesselmann, Phys. Rev. D53 (1996) 6638.

54. J. Fleischer and F. Jegerlehner, Phys. Rev. D23 (1981) 2001; O. Cheyette and M. Gaillard, Phys. Lett. B197 (1987) 205; W.J. Marciano and S. Willenbrock,

Phys. Rev. D37 (1988) 2509; S. Dawson and S. Willenbrock, Phys. Rev. D40 (1989) 2880; S.N. Gupta, J.M. Johnson, and W.W. Repko, Phys. Rev. D48 (1993) 2083; K. Riesselmann and S. Willenbrock, Phys. Rev. D55 (1997) 311.

55. A. Ghinculov, Phys. Lett. B337 (1994) 137; Erratum B346 (1994) 426; J. van der Bij and A. Ghinculov, Nucl. Phys. B436 (1995) 30; L. Durand, P.N. Maher, and K. Riesselmann, Phys. Rev. D48, 1061, 1084 (1993); Erratum D52 (1995) 553; A. Frink, B.A. Kniehl, D. Kreimer, and K. Riesselmann, Phys. Rev. D54 (1996) 4548.

56. J.F. Gunion, H.E. Haber, and J. Wudka, Phys. Rev. D43 (1991) 904.

57. H.E. Haber, G.L. Kane, and T. Sterling, Nucl. Phys. B161 (1979) 493.

58. W. Goldberger and M. Wise, Phys. Lett. B475 (2000) 275; G. Giudice, R. Rattazzi, and J. Wells, Nucl. Phys. B595 (2001) 250.

59. T. Appelquist and M.S. Chanowitz, Phys. Rev. Lett. 59 (1987) 2405; Erratum 60 (1988) 1589; F. Maltoni, J.M. Niczyporuk, and S. Willenbrock, Phys. Rev. D65 (2002) 033004.

60. J. Bagger, A. Falk, and M. Swartz, Phys. Rev. Lett. 84 (2000) 1385.

61. T. Appelquist and G.-H. Wu, Phys. Rev. D48 (1993) 3235.

62. Figure reprinted from: J. Erler and P. Langacker, *Electroweak Model and Constraints on New Physics*, in: K. Hagiwara et al. (Particle Data Group), Phys. Rev. D. 66 (2002) 010001, Copyright (2002) by the American Physical Society.

63. M.E. Peskin and J.D. Wells, Phys. Rev. D64 (2001) 093003.

64. J.A. Aguilar-Saavedra et al. (ECFA/DESY LC Physics Working Group), *TESLA: Technical design report. Part III: Physics at an e^+e^- Linear Collider*, DESY-2001-011, hep-ph/0106315.

65. K.J.F. Gaemers and G.J. Gounaris, Z. Phys. C1 (1979) 259; K. Hagiwara, K. Hikasa, R.D. Peccei, and D. Zeppenfeld, Nucl. Phys. B282 (1987) 253.

66. E. Accomando et al., *Physics with e^+e^- linear colliders*, Phys. Rept. 299 (1998) 1.

67. M. Diehl, O. Nachtmann, and F. Nagel, HD-THEP-02-34, hep-ph/0209229.

68. T. Barklow, in: American Linear Collider Working Group, *Linear Collider Physics Resource Book for Snowmass 2001*, SLAC-R-570, FERMILAB-Pub-01/058-E; see also: A. Para and H.E. Fisk (eds.), *Proceedings of the 5th Int. Linear Collider Workshop*, Fermilab, 2000, American Institute of Physics, 2001.

69. A. Pukhov et al., CompHEP, INP-MSU 98-41/542, hep-ph/9908288.

70. T. Stelzer and W.F. Long, MadGraph, Comput. Phys. Commun. 81 (1994) 357.

71. M. Moretti, T. Ohl, and J. Reuter, O'Mega, LC-TOOL-2001-040.

72. T. Ohl, VAMP, Comput. Phys. Commun. 120 (1999) 13; W. Kilian, WHIZARD, LC-TOOL-2000-039.

73. M.S. Chanowitz and M.K. Gaillard, Phys. Lett. B142 (1984) 85; G.L. Kane, W.W. Repko, and W.R. Rolnick, Phys. Lett. B148 (1984) 367; S. Dawson, Nucl. Phys. B249 (1985) 42; J. Lindfors, Z. Phys. C28 (1985) 427; J.F. Gunion, J. Kalinowski, and A. Tofighi-Niaki, Phys. Rev. Lett. 57 (1986) 2351.

74. E. Fermi, Z. Phys. 29 (1924) 315; E.J. Williams, Proc. Roy. Soc. A139 (1933) 163; Phys. Rev. 45 (1934) 729; C.F. von Weizsäcker, Z. Phys. 88 (1934) 612.

75. P. Garcia-Abia and W. Lohmann, EPJdirect C2 (2000) 1, and in [64].

76. P. Garcia-Abia, W. Lohmann, and A. Raspereza, LC-PHSM-2000-062, and in [64].

77. G. Gounaris, D. Schildknecht, and F. Renard, Phys. Lett. B83 (1979) 191, Erratum B89 (1980) 437; V. Barger, T. Han and R.J.N. Phillips, Phys. Rev. D38 (1988) 2766;

78. A. Djouadi, H.E. Haber, and P.M. Zerwas, Phys. Lett. B375 (1996) 203; A. Djouadi, W. Kilian, M. Muhlleitner, and P.M. Zerwas, Eur. Phys. J. C10 (1999) 27.

79. C. Castanier, P. Gay, P. Lutz, and J. Orloff, LC-PHSM-2000-061, and in [64].

80. F. Boudjema and E. Chopin, Z. Phys. C73 (1996) 85; V.A. Ilyin, A.E. Pukhov, Y. Kurihara, Y. Shimizu, and T. Kaneko, Phys. Rev. D54 (1996) 6717; V. Barger and T. Han, Mod. Phys. Lett. A5 (1990) 667.

81. K. Desch and N. Meyer, LC-PHSM-2000-052, and in [64].

82. E. Boos, H.-J. He, W. Kilian, A. Pukhov, C.-P. Yuan, and P.M. Zerwas, Phys. Rev. D57 (1998) 1553; Phys. Rev. D61 (2000) 077901.

83. V. Barger, K. Cheung, T. Han, and R.J.N. Phillips, Phys. Rev. D52 (1995) 3815.

84. T. Han, Int. J. Mod. Phys. A11 (1996) 1541; W. Kilian, Int. J. Mod. Phys. A15 (2000) 2387.

85. V. Barger, T. Han, and R.J.N. Phillips, Phys. Rev. D39 (1989) 146; S. Dawson, A. Likhoded, G. Valencia, and O. Yushchenko, in: *Snowmass 1996, New directions for high-energy physics*, hep-ph/9610299; T. Han, H.-J. He, and C.-P. Yuan, Phys. Lett. B422 (1998) 294; O. Eboli, M. Gonzalez-Garcia, and J. Mizukoshi Phys. Rev. D58 (1998) 034008.

86. R. Chierici, S. Rosati, and M. Kobel, LC-PHSM-2001-038, and in [64].

87. F. Larios, T. Tait, and C.-P. Yuan, Phys. Rev. D57 (1998) 3106; T. Han, Y.J. Kim, A. Likhoded, and G. Valencia, Nucl. Phys. B593 (2001) 415; J. Alcaraz and E. Ruiz Morales, Phys. Rev. Lett. 86 (2001) 3726.

88. Z. Kunszt, S. Moretti, and W.J. Stirling, Z. Phys. C74 (1997) 479.

89. For a recent review of Higgs physics at the LHC, see: D. Zeppenfeld, Int. J. Mod. Phys. A16 Suppl. 1B (2001) 835.

90. E.W.N. Glover and J.J. van der Bij, Nucl. Phys. B309 (1988) 282.

91. A. Djouadi, W. Kilian, M. Muhlleitner, and P.M. Zerwas, Eur. Phys. J. C10 (1999) 45; U. Baur, T. Plehn, and D. Rainwater, Phys. Rev. Lett. 89 (2002) 151801; Phys. Rev. D67 (2003) 033003; CERN-TH-2003-069, hep-ph/0304015.

92. D.A. Dicus, C. Kao and W. W. Repko, Phys. Rev. D36 (1987) 1570; E.W.N. Glover and J.J. van der Bij, Phys. Lett. B219 (1989) 488; Nucl. Phys. B321 (1989) 561; C. Kao and D.A. Dicus, Phys. Rev. D43 (1991) 1555.

93. V. Barger, T. Han, and R.J.N. Phillips, Phys. Rev. D38 (1988) 2766.

94. J. Bagger, V. Barger, K. Cheung, J. Gunion, T. Han, G.A. Ladinsky, R. Rosenfeld, and C.-P. Yuan, Phys. Rev. D49 (1994) 1246; Phys. Rev. D52 (1995) 3878.

95. A. Dobado, M.J. Herrero, J.R. Pelaez, E. Ruiz Morales, and M.T. Urdiales, Phys. Lett. B352 (1995) 400; A. Dobado and M.T. Urdiales, Z. Phys. C71 (1996) 659; A.S. Belyaev, O.J.P. Eboli, M.C. Gonzalez-Garcia, J.K. Mizukoshi, S.F. Novaes, and I. Zacharov, Phys. Rev. D59 (1999) 015022.

96. N. Arkani-Hamed, S. Dimopoulos, and G. Dvali, Phys. Lett. B429 (1998) 263; Phys. Rev. D59 (1999) 086004; I. Antoniadis, N. Arkani-Hamed, S. Dimopoulos, and G. Dvali, Phys. Lett. B436 (1998) 257.

97. T. Barklow, R. Sekhar Chivukula, J. Goldstein, Tao Han, et al., hep-ph/0201243, in: N. Graf (ed.), *Proceedings of Snowmass 2001*, SLAC-R-599.

Index

Springer Tracts in Modern Physics

Springer Tracts in Modern Physics